AND THE BAMBOO FLOWERS IN THE INDIAN FORESTS

And the Bamboo Flowers in the Indian Forests

What Did the Pulp and Paper Industry Do?

Volume II

MANORAMA SAVUR

IDPAD

MANOHAR
2003

First published 2003

ISBN 81-7304-413-9 (Set)
ISBN 81-7304-415-5 (Volume II)

Published by

Ajay Kumar Jain for
Manohar Publishers & Distributors
4753/23, Ansari Road, Daryaganj
Delhi 110 002

Typeset at

Kumud Print Service
Boileauganj, Shimla 171 005

Printed at

Rajkamal Electric Press
Delhi 110 033

Contents

Figures

Maps

Tables

CHAPTER 6

Karnataka

Why Eucalyptus and Why Rayon—Why not Bamboo and Why not Paper?

It's a land of heady spices,
A land where gold is found,
A land of scented trees,
A land of rich brown teak,
A land where the grass grows
into a mass of giant plumes,
It's the land where Bhadra flows.
Yes, the wise ones chose to make it
a land of milk and honey.
Attracted:
In hopped the innocent Hare
to nibble the bamboo shoots;
In romped the arrogant Bear,
to weave his wicked dream;
In sneaked the greedy Boar
to hold the foolish Hare's hand.
The silly Hare stretched its paw,
And lo, the Boar twisted its arm
and quietly nuzzled it out.
Now,
It's a land where the foolish ones hailed
The stately silver tree
And the green and gold acacia,
To salvage the bruised earth,
But it proved to be in vain
discovered the common farmer;
It's true, said the wise one;
I agree, echoed the keeper.
Yet, it's still a land of silver tree

A deceptive silver tree
And a deceptive green and gold.

M.S.

SECTION 1

Natural and Industrial History of the State of Karnataka

The British apportioned the Kannada-speaking area into the Mysore state, Madras Presidency, with a small bit added on to Bombay Presidency. The composite Karnataka state was formed in 1956. This study begins with the reconstruction of the natural history of its forests which began to change with the colonial interference in the system. Available literature, published accounts of early travellers, scholar-foresters like Stebbing, Gibson and Muthanna who chronicled events, have been used. Added to this are the various ecological assessments by scientists and foresters, besides critical analyses made by social scientists plus our team's own observations. From this, a changing picture of the forests of western Karnataka emerges.

The earliest accounts are recorded by Buchanan in 1801. Stebbing, a proud colonial forester, admitted that the destruction of what then appeared as a vast indestructible forest increased during the early days of British occupation. Occupation was initially to grab timber for ship-building, to keep the other European merchant companies at bay. He lamented that the British East India Company had no knowledge of tropical forestry, nor had European forestry as commercial forestry ceased to be understood as a commercial enterprise in Britain. Even after attaining supremacy over the country, the British continued to use wasteful forms of timber extraction for military purposes. There was no let up (Stebbing 1921: 34; Muthanna 1950: 400).

The Mysore wars by the British against the tenacious Tipu took great toll of the timber resources of the state. In the days of Tipu, a state monopoly had been created in respect of teak by declaring it a royal tree, so also with sandalwood (FRI, 1961). With the death of Tipu, Coorg and Malabar were ceded to the East India Company in 1779. 'Canara' was already in the hands of the British.

North Kanara, situated on the western ghats and on the west coast, was parcelled out as a part of the Bombay Presidency at that time. When the British gained control over Bombay state in 1818, they needed large quantities of hardwood for shipyards, military purposes, as well as for the construction of administrative and other buildings in urban centres. It is therefore not surprising that the British foresters were mainly interested in teak. In 1823 the forests were brought under the care of the Revenue Department. The British estimated that there was enough wood. The teak in North Kanara was of two types. Trees on the lower reaches, which were stunted, were used for building purposes and gun carriages. An estimated 1,54,000 trees were in existence. They were felled annually and supplied to the Mahrattas, fetching Rs. 5,424. The second type of tree, found on the ghats, was left untouched. British estimates put the number of these magnificent trees at 40,000. These were potentially useful for shipbuilding, but they survived the fate of felling due to their location as they were too far from any mountain streams to be floated down. Neither could they be stolen. Blane, the collector of Canara, and Gibson, the conservator in 1848, quoted authorities to show that the climate of the country was modified by rapid and extensive clearing away of the forests (Stebbing 1921: 119).

The state of Mysore was taken over by the British in 1832. This period witnessed the opening up of coffee plantations by clearing mountain slopes previously covered by moist forests (Muthanna 1950: 400). These coffee plantations were often abandoned after three or four crops and were reduced to a state little better than Kumri, or slash-and-burn agricultural areas (Stebbing 1921: 5).

Stebbing put down destruction of forests to Kumri. In 1848 Cleghorn, an intersted botanist, a trained surgeon and forest conservator by profession who had seen the destruction of forests in Shimoga, made a representation to the Governor to put an end to Kumri not only in Shimoga but in Kanara also. This was done. What is necessary to remember is that just 20 years later according to Brandis, large tracts of the country were again clothed with young forests (Stebbing 1921: 108). Light Kumri fire does not destroy the root system of teak and other hardwoods. But indiscriminate felling for military and other purposes definitely does, because when the canopy is opened up it paves the way for the aggressive *Bambusa arundinaceae* to colonize and flourish. In 1864, Shaw Stewart, the collector of North Kanara, who travelled through the forests of the region remarked on the wonderful luxuriance of bamboo (Stebbing 1921: 167). The British foresters insisted that colonization by bamboo was a result of Kumri and yet no one initially disturbed the forests as much as the colonizers did. This was around the time of the ban on shifting cultivation that coffee plantations were being promoted. One wonders whether the ban was meant to save the forests, or to provide labour for plantations (Gadgil 1989).

Stebbing boasts that the benign rule of the British and the security it offered led to increased agricultural prosperity and an enormous increase in flocks possessed by the community (Stebbing 1921: 296). The period he refers to is between 1796 and 1850 when India was beset with killer famines due to British rule and its neglect of agriculture. The conditions in Britain had changed with the introduction of the Spinning Jenny. Industrializing Britain was interested in making India its raw material producer of cotton, jute, tea, etc. Not only the large but even the smaller of the small ryots jumped into production of cash crops. To the small farmers a cow or a goat is a temporary insurance against starvation in bad times, particularly if cereal is given up for cash crops. The same Stebbing who holds that the enormous herds and flocks were responsible for the degradation of forests, forgets who was responsible for this upsurge in the population of livestock. Sadly, the Indian foresters also fail to see this and blame poor peasants whose economic situation continues to be bleak even today.

Meanwhile, in Mysore, after the temporary British takeover in 1831, Hunter noted that the forests were badly neglected. He made it his primary task to minimize destruction and unauthorized felling. Yet his budget for 1865-6 showed only Rs. 97,750 as expenditure on forest activities while the receipts were an enormous Rs. 3,15,719 (Muthanna 1950: 400). Brandis felt the forests of Mysore should be put in the care of a European-trained forester, and Van Somerson was picked for this task in 1866. He was a keen observer; he not only noticed that teak grew well in moist areas but also observed the local tradition of the villages of maintaining groves. He counted 16,000 *topes* (groves) in 12,138 villages. His first task was to survey the forest to mark out 'reserve areas'. With temporary rule coming to an end in 1881, the British felt that any expenditure on expanding forest activities was unjustified. Accordingly, the post of the Conservator was abolished in 1879 and only three trained officials were retained, the Deputy Commissioner being made responsible for the forest (Muthanna 1950: 400).

However, the favourable ecosystem was too tempting for the British to totally give up their activities. They turned to the setting up of plantations—initially revenue plantations of sandal, teak, gall nut and cardamom. Hutchison took great interest in introducing exotics like casuarina, eucalyptus, *Pitheoclobium sman*, *Saritonneia silgus*, *Toon*, *Sisoo*, mahogany, *Cera* rubber species, etc. Muthanna also notes that there was exchange of seeds between Bangalore and Brisbane (Muthanna 1950: 400). But Brandis, the Inspector General of Forests, did not approve of raising plantations on devastated forests. He firmly stated that plantation culture should be made subservient to the great work of restoring forests to at least something like normalcy by silvicultural means. But Brandis also wanted teak. He therefore conceived the notion of *toungya* which finally failed by the turn of the century (Stebbing 1921: 565-9). While still in service, Stebbing noticed that the second rotation crop of teak, though a native species, did not yield as much as it should. He also warned that it was a mistake to remove all the undergrowth of the plantation as was consistently done. With such a practice the ground turns hard and dry with little humus (Stebbing 1921: 430).

The first working plans for the Mysore state were made under the guidance of Pigot in 1899. By the Mysore Forest Regulation Act of 1900, the Indian Forest Service was established and Muthanna became its head. He had the forests surveyed and found a surprising reduction in the area of reserve forests (Muthanna 1950: 402). The next damaging shock to the forest came in the form of the two wars. That the Mysore state was not a part of British India was irrelevant. In a short period of two years, 1917 and 1918, 30,000 sleepers, wood for railway tracks from hardwood trees and 90,000 pegs for tents were sent out. For the Second World War too, immense timber was supplied. During the peak period 1943-4, 8,00,000 cu ft of logs, 2,765 transmission poles and 60,000 bales were harvested from the forest. The sawn material alone was more than 270,000 ft. These were distressing years for the forest (Muthanna 1950: 403).

Meanwhile, from 1905 onwards, successful research on the use of bamboo as pulpable material by Raitt at FRI yielded good results. But the question of actual production was intercepted by the First World War. Nevertheless, Raitt's interest in bamboo forests persisted and two officers, H.S. Narayan Rao and G. Krishnamoorty Naidu, were placed on special duty to survey the forest resources of the state. The British government was still embroiled in the war. The war was followed by a short boom and then a depression. Though during the period of the post-War boom, Britain made an attempt to manufacture paper and hold a monopoly control over it through various tariff controls, there is no written literature to show that it had interests in the bamboo forests of Karnataka.

Balgi poles were used for electrification, the Mysore Match Factory was revived, a plywood factory was established in Bangalore, and most important, the FRI laboratory was established in 1938 at Bangalore (Muthanna 1950).

The present-day forests of Karnataka comprise:

1. Forests of Bellary district of Madras state which merged with Mysore in 1953,
2. Forests of Coorg merged in 1956,
3. Forests of North Kanara formerly in Bombay state merged in 1956.
4. Forests of Kollegal divisions in the northern part of Coimbatore and the forests of South Kanara district of Madras merged in 1956.
5. Forests of the former princely state of Mysore.

Political History of Mysore

It is necessary to go backwards in time to understand the relevant situation. In the year 1800 Mysore came under British suzerainty. True, the internecine wars and the war with the British had ended. But these had meant a direct drain of Rs. 245 lakh to the British crown, and had also repeated the serious food scarcity and near-famine conditions for the people. This happened in 1824, 1831, 1833, 1850, 1865 and 1875. British troops were stationed in the Bangalore cantonment. Troops movements and movement of food grains necessitated the building of railways. In 1864 a railway link was established with Madras which was a British stronghold.

Mysore was rich in minerals, forests and water resources. But the British cared to open up only the Kolar gold fields and permitted the rich variety of cottage industries found throughout the state to almost die out. The British withdrew in 1870. This gave the state an oportunity to rise again, and to rise vigorously.

Industrial Development of Mysore State

The new rulers of Mysore, Raja Wodeyar III and Chamraja Wodeyar, patronized learning and the arts. Meanwhile, the forward-looking Dewan K. Sheshadri Iyer (1883-1901) set development in motion. He set up the basic infrastructure such as hydel power in 1900, as well as railways, which attracted Bombay firms to establish spinning and manufacture of cotton textiles in Davanagere and Bangalore. The next progressive step was state planning undertaken by Sir (Dr) M. Vishwesvarayya, the engineer statesman, the next Dewan of Mysore from 1912-18. His visible contributions were in the fields of water and electricity, laying the infrastructure for subsequent industrial growth. He was responsible for initiating and overseeing the development of iron and steel works at Bhadravati in the 1920s. His most important contributions were intangible; he instilled a vision and a measure of self-confidence among the people. The next illustrious Dewan was Sir Mirza Ismail who served twice, once in the late 1920s and again for a longer term through most of the 1930s. Mirza is credited with having developed the pattern of state entrepreneurship in industry which was distinctive of Mysore's pre-war development. He started an impressive array of basic and consumer industries with state capital at a time when private capital was still shy and entrepreneurship not forthcoming (Ramchandra 1961). Entrepreneurship was also encouraged through financial aid to set up what is today commonly known as the joint sector. Together, the public and joint sector included iron and steel, cement, soap, chemicals, fertilizers, silk, cotton textiles, porcelain, electricity and paper, apart from traditional industries. These were set up before the Second World War. Mirza also attempted to revive traditional industries.

In the quotation which follows, the bracketed information is that of the study team, ST.O. = State-Owned, and ST.A. = State-Aided, Inv. = Investment. In a broadcast in 1938 Diwan Mirza Ismail quoted with pride:

We Are Very Proud Of Our Factories,
And At The Risk Of Being Called Provincial
Try To Set Before All True Mysoreans
The Ideal That They Should Wash Themselves
With Mysore Soap (1918 ST.O. Inv. Rs. 2.31 lakhs)
Dry Themselves With Mysore Towel (1920 Inv. Rs. 1.92 lakhs)
Clothe Themselves In Mysore Silks (1932 ST.O. Rs. 4.2 lakhs)
(1938 ST.A. Rs. 10 lakhs)
Ride Mysore Horses,
Eat Abundant Mysore Food (OIL, 1930)
Drink Mysore Coffee With Mysore Sugar (coffee ST.A. sugar 1938 ST.A.)
Build Their Houses With Mysore Cement (1936 ST.O.)
Mysore Timber And Mysore Steel (1918 ST.O.)
Furnish Their Houses With Mysore Furniture (1913 ST.O.)
And Write Their Letters On Mysore Paper (1936 ST.A.)

Source: 1938, Mysore and Modern Industries.

It is important to understand how Mysore—a native state—decided to set up a paper mill in the mid-thirties, when such an industry was not permitted by the British. The mill was set up in Shimoga.

By the 1930s the political situation was changing. With the crisis of World capitalism and waning of British power the Indian Industrialists in the eastern part of the country—the Birlas, the Dalmias, the Singhanias, the Bajorias—were pushing forward and by the mid-1930s had penetrated into the British stronghold—the paper industry—which it zealously guarded. The so-called native states were not to be left behind. Mysore was one of them, and the Nizam of Hyderabad was the other.

Vishwesvarayya, the person to be credited for the industrial development of Mysore state, showed interest and called a conference of the forest officers of the state. As explained elsewhere, the British monopoly control over the manufacture of paper was broken by four industrial houses and by the Maharaja of Mysore when the British stranglehold over India was weakening. In 1937 the Mysore Paper Mills was established in Bhadravati. The infrastructure for industrial production had already been laid in this town for the commissioning of the charcoal-based Bhadravati Iron Works. 1935 onwards was, for Mysore, a period of prosperity. Shimoga and its surrounding areas were earlier thick jungles, abounding in bamboos. This attracted the attention of Raitt, the discoverer of the pulpable qualities of bamboo, and a survey was carried out. Was it to set up a British-controlled paper mill? This is not known. River Bhadra flows through Shimoga, the area falling in the Maharaja of Mysore's territory. And Mysore had won a reputation as a 'progressive and energetic' state. Bhadravati, in Shimoga became a paper town.

SECTION 2 (A)

Locational Status of Forests of Uttar Kannada and Shimoga

The three pulp and paper industries of Karnataka and their catchment areas are located in the western region of the state. The concern of this study is with the forests in this region. While the West Coast Paper Mills is located in Uttar Kannada, Mysore Paper Mills is located in Bhadravati in Shimoga district. The third unit, Harihar Polyfibres, is

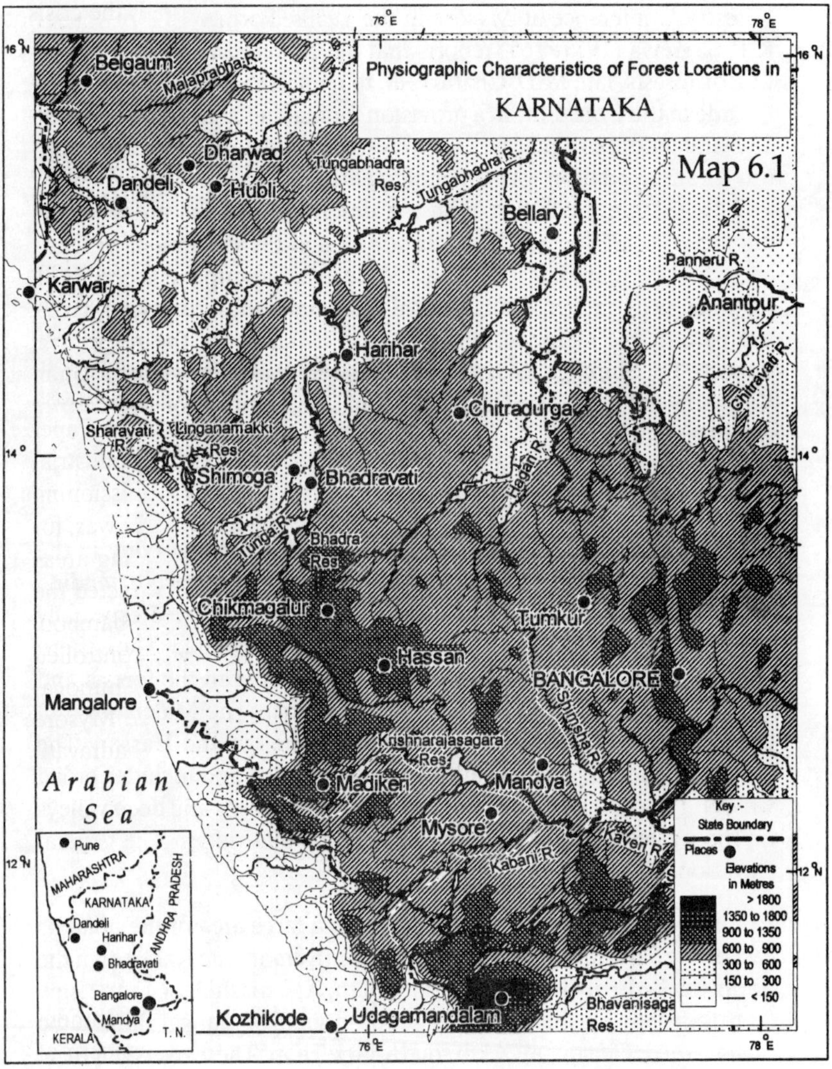
Physiographic Characteristics of Forest Locations in
KARNATAKA
Map 6.1
Belgaum
Malaprabha R.
Dharwad
Dandeli
Hubli
Tungabhadra
Res
Tungabhadra R.
Bellary
Panneru R.
Karwar
Varada R.
Anantpur
Harihar
Chitravati R.
Chitradurga
Sharavati R.
Linganamakki
Res
Shimoga
Bhadravati
Hagari R.
Tunga R.
Bhadra
Res
Chikmagalur
Tumkur
Hassan
BANGALORE
Mangalore
Shimsha R.
Krishnarajasagara
Res
Madikeri
Mandya
Arabian
Sea
Mysore
Kaveri R.
Kabani R.
Key :-
State Boundary
Places
Elevations
in Metres
> 1800
1350 to 1800
900 to 1350
600 to 900
300 to 600
150 to 300
< 150
Pune
MAHARASHTRA
KARNATAKA
ANDHRA PRADESH
Dandeli
Harihar
Bhadravati
Bangalore
Mandya
KERALA
T. N.
Kozhikode
Udagamandalam
Bhavanisaga
Res
16° N
14°
12° N
76° E
78° E

located in Harihar. Western India plywood is outside the scope of the study. The drain on the forests by the industry being heavy, and the foresters deeply concerned, the issue was discussed at the Asia Pacific Forestry Conference at Mysore in the 1950s. Former I.G. of forests K.P. Sagreiya (1977: 165) reports that the conference 'authorised the I.G. of forests, late M.D. Chaturvedi, to request the Prime Minister to include in the Constitution a provision for making it illegal to disforest any land unless an equal area was forested satisfactorily. Unfortunately this recommendation was not implemented.

This section deals first with the description of the areas where the factories are located, the location of and the catchment or lease areas for raw materials till half a decade ago. Descriptive accounts of parameters like relief, drainage, soil types, rainfall, forest types, land-use pattern are recorded by the *District Gazetteers* of the respective districts. For Shimoga, *Gazetteers* of 1985 were available, while for Uttar Kannada the team used the 1883 and 1985 edition of the *District Gazetteers* only.

Uttar Kannada District (UK)

Relief: The western ghats divide UK into two; one half lying in the coastal plains and the other above the ghats with some parts extending into the Deccan Plateau. The district is hilly and wooded especially in the highlands. The ghat, the Sahyadri region, is forested. The eastern margin is undulating. It is a transition zone between the forests and agriculture lands of the upper Dharwar district. The Malenadu (hill country) of the Sahyadris continues into Shimoga and Hassan. The eastern portion of the Sahyadris is more rounded and includes broad valleys. The central portion consists of crestline ridges and deep valleys on the western side, the hills rising in spurs and ridges. The general elevation of the area is 675 m.

Drainage: Large rivers in the district drain a large area of the uplands east of the Sahyadri scarp. The four major rivers are the Kalinadi in the north, the Bedti or Gangavalli 32 km south, Aghanashini or Tadri rising far to the south, but falling into the sea 6 km south of Gangavalli, and the Sharavathi or Gersoppa, 24 km south of the Tadri. These rivers originate in the western ghats traversing the district, and meeting the Arabian Sea.

Rainfall: The average rainfall is 2,741 mm. Heavy rainfall is received during June and August. Over the years no large scale variations have been noted.

Soils: Lateritic soils are dominant in this region, alluvial soils being found along the rivers. Black and red soils are also a common feature. The entire western slopes of the ghats have fairly deep soils. On the lower lands the soil is black with an underlayer of red. Where teak is abundant, the soil is lighter in colour, loose and mixed with quartz.

Forests: Before Karnataka came under British rule, its forests supplied the shipbuilding yards of Haidar Ali (1761-82) with fine teak and other timber.

In 1883 the forests were entirely the property of the British government, certain privileges being allowed in protected forests. The forests were found in three areas above the Sahyadris, along the main range of the Sahyadri and on the western spurs of the Sahyadri. Central Kanara had plateaus which were thickly wooded. The higher slopes were hidden by magnificent forests where the vegetation was temperate. Towards the east the forests dwindled into stunted teak and scrub type.

While the *Gazetteer* of 1883 stated that there were patches of evergreen forests along the slopes, the majority of the forests were of the leaf-shedding type. The Sahyadris were clothed in evergreen forests. Bamboos were reported to be in abundance.

The 1985 *Gazetteer* classifies the forests as being of the:

(1) Evergreen type, found in a narrow strip along the ghats where large canes are present, poor in grasses and relatively few orchids.
(2) Semi-evergreen forests found near the evergreen forests.
(3) Moist deciduous forests found along the western coast and eastern uplands.
(4) Scrubs and thorny forests of the plains.
(5) Unwooded forests.

Land Use: The valleys of the ghats sustain human activity as seen in the belts of spice and areca gardens for which the district is famous.

The *Gazetteers* of 1883 give accounts of coconuts and rice lands along the coast, valleys whose basins are crowded with spice and betel gardens and stream banks where spice, rice and sugar cane were grown in the uplands.

A highly eroded and rocky landscape is the most striking aspect of the UK coast. The soils in the ghat regions are loose and soft and vulnerable to erosion if the plant cover is lost. The superficial lateritic deposits are of recent origin.

M.D. Subash Chandran (1984), says of the vegetation of the area, that it improves in density and quality in the higher terraces where some of the best teak stands can be seen. Several biotic factors have reduced the extent of the forests. Cleared forest lands have been taken over by the weed, *Chromolaena odorata*. When the ecological conditions of the evergreen forests are modified, semi-evergreen forests are formed. When the canopy is opened up, species from drier zones invade. *Callicarpa tomentosa*, *Macaranga peltata*, *Trema orientalis* are the early colonizers of disturbed areas; the original trees are replaced by *Terminalia spp. Bombax spp.* And typical deciduous trees (*Gazetteer*, Uttar Kannada 1985). Thus semi-evergreen gives way to deciduous, which seems to have been the case here too, since the deciduous forests are found surrounding the semi-evergreen.

Over the years, increasing agricultural activities have also contributed to changes in the overall landscape.

Shimoga

Relief: Shimoga, located in the mid south-west part of the state, consists of mountainous terrain (*malenadu*). In the eastern region there is a transition from *malenadu* to semi-*malenadu* to *maidan* (plains). The greater part of the district lies in the *malenadu*. The open country is devoid of rich forests. The district slopes from west to east, the general elevation of the area being about 640 m.

Drainage: The important rivers are the Tunga, Bhadra, Tungabhadra, Sharavathi, Kumudvati and Varada.

Forests: According to the 1971 *District Gazetteer*:

1. The evergreen belt is confined to the western part of the district. Many of the hills are covered with heavy forests; common species being *Vateria indica*, *Artocarpus hirsuta*, *Calophyllum elatum*, *Hopea parviflora*.
2. The moist deciduous belt is found in the extreme north of Sorab taluk and extends southwards. It includes timber yielding forests and much sandalwood. The prominent species are *Tectona grandis*, *Dalbergia latifolia*, *Pterocarpus marsupium*, *Terminalia species*, *Santalum alba*, etc.
3. The dry deciduous belt—the tree growth is inferior and of much smaller size. The common species are *Anogeissus latifolia*, *Azadirachta indica*, *Terminalia chebula*, *Zizyphus species*, *Acacia nilotica*, etc.

The forests covered nearly 20.5 per cent of the land in the district in 1971. The moist deciduous is the dominant forest type.

Impact of the Industry in the Catchment Area

This section attempts to present a picture of the catchment areas of the paper and pulp industries of Karnataka, viz., Mysore Paper Mills Ltd. (MPM) and West Coast Paper Mills Ltd. (WCPM) and Harihar Polyfibre, now known as Grasim in Harihar. Numerous studies made by Madhav Gadgil and Narendra Prasad throw light on the status of their catchment areas for the period between 1960-70, and also explain the reasons for the depletion of the natural resources. The year 1990 saw the Government of Karnataka issue orders that put an end to felling in natural to areas and also to raising of plantations. Our own surveys in 1993 in the catchment areas highlight the status of the resource base in the wake of the government action.

MPM was incorporated in 1937 and before doing so, a survey of the bamboo resources of the state was made. Kadambi (1949) states that in 1935, C.H. Eagle's paper on 'The Probable Annual Yield of Bamboos in the Bhadravati and Shimoga Districts', noted that 22,537 and 25,860 tonnes per annum respectively could be sustainably extracted from these areas. Another estimate by M.G. Venkatrao in 1937 stated that 16,920 tonnes could be extracted sustainably. The Government of Mysore had to base its estimates on these widely varying accounts. Kadambi pointed out that these varying estimates were the result of differing opinions as to how many culms make a tonne. According to Gadgil and Prasad (1978) the 40-year old working plans of the Mysore state prescribed the eradication of bamboo as a weed in teak plantations. The working plans also stated that there were 'unlimited stocks' of bamboo. Again, in making their estimates, it was not taken into account that bamboos flower and die gregariously. Therefore these estimates of sustainable yields were over-estimates by an almost ten-fold factor (Gadgil 1991). Gadgil thinks that there were 'pressures' on the Forest Department (FD) to present such overestimated figures. One can only guess as to who stood to gain by these exaggerations.

In 1937 the Government of Mysore conceded the entire forest areas of Shimoga and Bhadravati to MPM for the extraction of Bamboo (Gadgil 1978). After independence when India launched a massive drive to industrialize, Uttar Kannada (UK) became a major centre for forest-based industries. State policies called for encouraging industry in every way by guaranteeing the supply of raw materials on a long-term basis at nominal rates. Reserve forests were given over on lease to industries for the extraction of timbers. 1946 saw the Indian Plywoods, owned by Kadar Kutty, a large private company, being set up in UK (see Ch. 7—Kerala, for growth of Western Indian Plywoods). WCPM in 1958 was the second paper mill, which was set up at Dandeli. The mill was accorded certain concessional areas in UK and adjacent regions. This period also witnessed an explosion in population in the UK region due to immigration by those attracted by employment opportunities generated by the new industries as also by the upcoming Kali river hydroelectric project, and an increased demand on forest resources by the industrial and urban sectors.

Meanwhile at Bhadravati much of the bamboo stock flowered and died in the mid-1950s, following which MPM started exploiting bamboo from Karwar, Mercara, Mysore, Chamarajnagar and Kollegal divisions. Regeneration which usually takes 10-15 years, failed, but MPM continued to extract from these and also from Mangalore and Coondapur divisions (Gadgil and Prasad 1978). In UK, too, the bamboo began to flower gregariously in 1959. Since bamboo was gregariously flowering in waves, the industry began harvesting from various districts like Yellapur, Sirsi, Karwar, etc., and collecting dead and dry bamboo. When bamboos flower, the seeds are dispersed and the area needs protection for at least a year to help the bamboo seedlings to establish themselves. The foresters believe that young seedlings faced danger due to the increased pressure of cattle grazing and the light fires spread by the local communities (WCPM, official personal communication, 1993). WCPM, in a bid to save the upcoming bamboo, in 1964 proposed to the government that they would protect the areas by putting up three rows of barbed wire fence around 60,000 acres of forest and by under-planting bamboo in some 48,500 acres within this area and by undertaking rigid fire protection measures. The government agreed to this. When the mill authorities fenced this extensive forest, the cattle around were deprived of their grazing lands. In 1972 the villagers agitated for the right to graze cattle in the forests. The state government ordered barbed wire fencing to be removed to allow village cattle to graze. The representative of the mill claims that 8-year bamboo clumps were totally destroyed. This however, was not borne out by the team's earlier observation in December 1985.

It is interesting to note here that in 1960 the market price for bamboo was around Rs. 3,000 per tonne and the government policy to subsidize raw material saw to it that the industry was awarded bamboo at Rs. 3.12 per tonne of paper, which works out to Rs. 1.5 per tonne of bamboo (Gadgil 1989).

WCPM foresaw the crisis in raw material and took up its own plantations. In 1962 they acquired 5,700 acres of some private *inam* lands in Dharwad district at an annual rent of Rs. 9.50 per acre for a period of 30 years. These lands were slopes which may have supported forests not too long ago. The company raised 100 acres of *Eucalyptus citriodora* and eucalyptus hybrid on the remaining area. They claim that they were the pioneers in raising industrial plantations with exotics (WCPM, official personal communication, 1993). Their plantations were highly successful. They tried numerous experiments by interplanting bamboo with hemp, *kenaf*, etc. (all good pulp yielding plants), as well as inducing farmers to grow *kenaf* by providing them with all necessary inputs (personal communication, December 1985). It is not clear when and why the mill stopped this arrangement. The company also acquired some 3,000 acres of *inam* lands in Belgaum district where they raised bamboo successfully. WCPM experiments on private lands initially yielded good results and may be held up as an example of what companies can do to meet their raw material demand.

In 1965 Karnataka saw the third paper mill, Mandya National Paper Mill, being set up. MNPM was promised 1,40,000 tonnes a year of bamboo but in fact it received not more than 25,000 tonnes per year (Gadgil 1978). Mandya was soon to go out of production, till in 1986, the Public Sector Unit, Hindustan Paper Corporation, was compelled to take it up as a sick unit and nurse it back to health. It was converted into a bagasse based unit, and therefore falls outside the scope of this study.

But sometime in 1972-3 there were protests by the local people against the barbed wire fences put up by the WCPM and the trench it had dug around the forest like a moat. The government relented and WCPM was asked to remove the fences. The mill authorities say that in disgust they gave up any further activities in promoting regeneration of the areas. The team's observations of the area in December 1985 revealed that only one wire of the fence was removed. The other two wires were left untouched, the fences now being interspersed with small wicket gates manned by watchmen. The trench, a mini-moat, had not been filled up. WCPM's attempts to ensure the regeneration of bamboo in the forests were considered successful (Gadgil and Prasad 1978).

Meanwhile in the 1970s, MPM was struggling to meet its annual requirement of 55,000 tonnes, being able to get only 25,000 tonnes as against the promised 1,44,000 tonnes per year. Sustained yields for the concessional areas allotted to WCPM were estimated to be 1,50,000 tonnes per year, but the actual annual yield was only 40,000 tonnes per year. The period from 1960-80 saw industrial demand peaking along with the natural resource supply increasing (Gadgil and Prasad 1978). Till 1975 their raw material requirement was met through the flowering bamboo, after which the demand began to be met from less accessible areas. This industrial demand saw to the rapid decimation of the resource base. Surveys by Gadgil and Prasad (1978) report that areas which were now entirely lacking in bamboo had been extremely rich in bamboo earlier. Gadgil (1989), investigated two prevalent hypotheses put forth by the foresters that it was the grazing pressures and the fires set off by the local population that prevented bamboo regeneration, and secondly that of the basket-weavers. Gadgil and Prasad hold that it was the harvesting pressure of MPM and WCPM that prevented the regeneration of the bamboo. This is also our hypothesis and is confirmed by our observation in eight other states.

Investigations by Gadgil and Prasad revealed that in their effort to meet the raw material demand in the face of scarcity, the contractors hired by the two paper mills disregarded silvicultural rules. They extracted material from the accessible areas, nearer to the roads, consequently, certain clumps were over-exploited. The labourers hired to fell the bamboos could not recognize the age of the culms. Again, the clearing of the thorny cover at the base of the clump as was practised left the young shoots unprotected and exposed to the grazers (Prasad 1985). While the local population may extract one or two culms at a time, the company extracts hundreds.

The environmental consequences of the depletion of bamboo has been discussed by Gadgil and Prasad (1978). Ecologically, bamboo is a sun-loving pioneer species which can grow profusely in moist and dry deciduous forests wherever the canopy has been opened up. When bamboo disappears this niche is taken over by *Eupatorium glandulosum*. This weed totally suppresses all other regeneration. It dries up in summer, promotes fires, causing loss of valuable mineral resources which could have been conserved by bamboo. This weed also has certain allelopathic chemicals which adversely affect the growth of other plants. It cannot be grazed upon and neither does it support any wildlife. Therefore the growth of *Eupatorium* over vast tracts in Karnataka has serious deleterious environmental consequences.

In 1978, the Karnataka government put a ceiling on the size of landholdings and lands under the Karnataka Land Reforms Act, wherein the lands became government property. This led to a battle over ownership of the land earlier acquired by WCPM. The leases were valid till 1992. Besides these private lands, the industry had been given 298 acres of degraded land at Honnapur near Kulawalli by the FD in 1981, on which the company raised bamboo and some eucalyptus. These are very steep lands with lateritic soil. The yield from these lands, says WCPM, has never been harvested because the FD demands royalty on the yield and the company refuses to pay since they have borne the cost of raising these plantations apart from paying the rent. WCPM argues that the royalty was not mentioned in the original lease agreements, and that the government is probably demanding it in the face of the new restrictions on forest produce according to the 1980 Forest Policy.

By the mid-1970s bamboo was becoming a limited resource. Other sources had to be tapped. The decline in the naturally available raw material was sought to be solved by promoting large-scale plantations with exotics. It was in 1976 that the National Commission on Agriculture issued a directive supported by special funds provided by the five-year plans to clear fell natural areas in order to grow fast-growing species. This has been discussed in Chapters 3 and 7. Gadgil (1989), notes that no proper scientific research went into the establishment of these plantations as a result of which they were failures, especially those raised on the high rainfall tracts in UK district. The few remaining eucalyptus trees in Dandeli stand as silent witnesses to the lack of research on the species—site relationship, which has led to attacks of the 'pink disease' on the eucalyptus in this area (ACF, Haliyal division).

The large scale exploitation of natural forests and the massive 'afforestation' monocultures which proved a failure, saw the government of Karnataka issue two orders after the Forest Conservation Policy of 1980. Outlined below are the salient features of these orders. According to G.O. No. 164 dated 10 October 1990, the existing total ban on felling in any natural area would be allowed to continue for a period of three years, with absolutely no felling allowed in evergreen areas. Bamboo and cane were to be extracted only for the artisan community. The wood from the artificially raised plantations could be harvested. According to the G.O. No. AHFF 164 EDP 90 dated 10 October 1990 there would be a ban on supply of raw material from FD plantations and natural areas. No fresh leases on forest lands would be issued, nor would expired leases be renewed. The forest produce extracted would be transported to government depots and sold at public auctions.

In the wake of these orders the companies no longer extract bamboo from the forests. Even earlier expired leases for the extraction of bamboo were not renewed. The Forest Department extracts the dried up and dead bamboo and sells it at an open auction where the companies may buy it.

Our own surveys in 1993 revealed that bamboo was coming up all over the districts. Young clumps were observed to be growing on the hill slopes adjoining the pulpwood plantations in Bhadravati. It may be noted here that these hill slopes were not cleared of their natural vegetation as they were inaccessible to the bulldozers. But labourers were hired to work in the area.

Areas around the MPM mill, part of the old catchment area reveal barren hill slopes with sparse vegetation of *Zizyphus spp.*, *Solanum spp.*, indicating the presence of the scrub jungle. There are patches of dry deciduous vegetation with *Dendrocalamus strictus* being dominant. The extent of moist deciduous patches was small as also that of dry deciduous vegetation. Scrub jungles dominate the scene, indicating a highly degraded habitat. Queries made to the local people revealed that there had been more forests in the past with a lot of wild animals and a lot of *Matti* (*Terminalia* species), *Beeta* (*Dalbergia* species) and sandalwood trees.

MPM gets supplies of eucalyptus and acacia from its own captive plantations, raised on 30,000 ha of land in Shimoga and Chikmagalur leased from the FD in 1980 for a period of 40 years. Gadgil (1989) reports that these lands were earlier marked for raising biomass to meet the local needs. MPM also gets eucalyptus and acacia from the farmers of Tumkur and Kolar. Till 1990, only 25 per cent of the bamboo requirements were met from the natural forests. The industry imported 570 tonnes from local farmers and 37,800 tonnes from neighbouring states (Lakshmana 1991). An executive of the MPM factory said that bamboo was also being supplied from homesteads in Kerala.

All the companies continued to extract recklessly from the areas leased out to them till the late eighties. At WCPM the leases expired in 1992 and have not been renewed. The company gets only a minimal quantity of bamboo from the Forest Department. Even earlier, as stocks began to decline, WCPM was forced to go further and further away, from Andhra Pradesh to Garhwal to Assam to Nagaland and

Meghalaya in search of raw material (Gadgil 1989). But bringing bamboo from such far away areas was uneconomical. Currently WCPM uses 55 per cent eucalyptus, 25 per cent casuarina, 10 per cent *Subabul*, 6 per cent bamboo and 4 per cent of *Erythrina superosa*. The eucalyptus it gets from the Karnataka Forest Development Corporation (KFDC) and private farmers of Karnataka, casuarina from Tamil Nadu, *Subabul* from Andhra Pradesh, bamboo from Haliyal division and the *Erythrina* from private sources in Karnataka.

Our surveys of the forests of Haliyal division in October 1993 (a part of the old catchment area of WCPM) revealed *Bambusa arundinacea* growing in very dense patches. The dominant forest type is of the moist deciduous type. Large areas covered with plantations of teak interspersed with pockets of natural vegetation, standing as reminders of what the earlier vegetation may have been, were cleared to raise these plantations. The industry complains that the Forest Department does not extract bamboo to its fullest potential. The FD allows the artisan community to remove only one to two culms per day, after getting a certificate from the DFO concerned. What is interesting is that bamboos have began to flourish again, felling by the artisans apparently having no adverse effect on its growth. This counters the general impression that bamboo forests react to the pressures from the common people, particularly bamboo workers. The FD is also involved in raising bamboo plantations for the artisan community. The large scale decimation of the forests that is seen in Dandeli can be attributed to the initial clear felling resorted to by the colonizers to raise teak, the loss incurred when large areas got submerged under the Kali river hydel project, and the demand from the plywood and pulpwood industries in this area. Evergreen forest patches can be seen in isolated pockets. The luxuriant bamboo growth visible now can largely be credited to the halting of all bamboo extractions since 1989. Bamboo continues to be extracted but only by the local communities, and to a certain degree by the State Forest Department but this is confined to the dead and dry bamboo.

While the dispute over the ownership of the private lands acquired by WCPM continued, the WCPM carried out all plantation activities by paying the rent to the Forest Department. This was possible only till 1992 which was the final year of the lease period. The company extracts all the available eucalyptus, but when it is time for replanting, they are not allowed to carry out since a verdict has not been reached on the ownership issue. The well-tended Kulawali plantations that we saw in December 1985 now stand as barren hill slopes in 1993 (October).

The few thin coppices of eucalyptus that remain are being extracted by the company. Our observations in 1993 revealed vast areas covered by a weed—*Eupatorium*. In some areas labourers were hired to clear away the undergrowth. In areas not taken over by *Eupatorium*, young trees of teak, *Terminalia* species, and even a few saplings of *Dalbergia* species were observed to be coming up. Natural vegetation was also seen sprouting, since no new eucalyptus plantations were being raised. This coincides with the observation made in Kerala that grasslands, after felling *Eucalyptus grandis* plantation affords hope for reversal to natural forests.

On the raw material issue the industries were very comfortable in early 1993 as there was a temporary glut of paper in the market. Earlier the GOK had issued prohibitory orders on the transport of certain timber species. Even eucalyptus fell into this category. According to these orders, transporting of such timber would need special permits. This restriction has been lifted since 1992 and new farmers are now cutting down as much as they can.

Augmenting Forest Raw Material for PPI by KFD, KFDC, KPP

In this section the area under plantations is dealt with, as is the species cultivated, and the investment made by the Forest Department on its own forest lands in Karnataka. The three plantation corporations are highlighted. These are the Karnataka Forest Development Corporation (KFDC), the second one being the Karnataka Pulpwood Plantation, a joint sector one for rayon industry (HPF) presently under public litigation. This is a fair indication of the mood of the people. The third one is the corporation to which the Forest Department extends technical assistance and leases out its land. This lessee is a joint sector company MPM, which now produces newsprint and some amount of writing and printing paper.

Incidentally, the pulpable species are also grown as far as possible on all available public lands as part of the social forestry programmes, and in some private lands as farm forestry. Karnataka, like other southern states, is turning green with a large component of exotic pulpable species, especially eucalyptus and *Acacia auriculiformis*.

Incidentally, since the end of the nineteenth century, the Mysore state was involved in raising plantations of fast-growing species but in a small way.

It was FAO which formally suggested eucalyptus as a pulpable species as late as in 1961, though it had been working on this since 1949 (FAO, 1947). Interestingly, the first meeting, where Von Monroy was present, of FAO was held in Mysore as early as in 1949 (Huberman and Purukayastha 1951). It perhaps explains why eucalyptus was introduced quietly in forestry in 1955 (a retired CCF, personal communication, September 1994).

Post Colonial Forestry: Pulpwood Plantations

Though the Karnatka Forest Department (FD) has been involved in raising eucalyptus since 1950, this was not quite enough to meet the industrial demand (Chaturvedi 1976: 58). This industrial demand for hardwood came up during the 1960s when overfelling resulted in a decline in the bamboo stocks of the state.

The year 1971 saw the Government of Karnataka form the Karnataka Forest Plantation Corporation (KFPC) with the object of raising industrial plantations like eucalyptus, cocoa, rubber, etc. Incidentally, mention is also made of other types of plantations raised in forest areas since we consider monoculture as detrimental to the forest ecosystem.

The paper industry faced a grave shortage of natural raw material and in 1976, the National Commission on Agriculture recommended promotion of plantations (discussed in detail in earlier section).

By 1977-8, KFPC had raised 49,920 acres of eucalyptus with loans from the Agriculture Refinance and Development Corporation (ARDF) for an amount of Rs. 137 lakh. KFPC raised its plantations on land leased out by the FD. The Forest Department conceded the eucalyptus plantations raised by KFPC to MPM, Harihar Polyfibres (HPF) and HPMC—Mandya for five years at Rs. 65 per tonne. In 1977-8 the corporation earned Rs. 106.00 lakh from its plantations leased out to HPF and WCPM. Besides eucalyptus, the KFPC had plans to raise 5,000 ha of cocoa. Also, the FD transferred its rubber plantations to KFPC (KFPC, *Annual Reports*).

It was in 1978 again, that the Karnataka Forest Department decided that instead of growing industrial raw material in the natural forest, at a cost where the industry would benefit, the industry should instead be offered degraded reserve forests to raise its own captive plantations. The Karnataka government accepted this proposal. Its definition of degraded forests was Reserve Forests with less than 2 tonnes of trees

TABLE 6.1: SPECIES-WISE AREA UNDER PLANTATION BY MPM

(*in ha*)

Year	Eucalyptus hybrid	Eucalyptus camaldulensis	Acacia	Pinus	Casuarina
Upto 1989	6,549	521	7,392	817	1,387
1990	0	806	909	0	0
1991	0	744	517	364	0
1992	0	500	688	200	0
1993	0	0	1,281	1,000	0

Source: MPM's Forestry Wing.
Note: The industry proposes to stop eucalyptus plantations and raise only acacia and pinus.

grown per acre (Shyam Sundar and Parameswarappa 1987: 192). Only MPM accepted this offer.

In 1980 the state government conceded 30,000 ha of D-Class lands for a period of 40 years in Shimoga and Chikmagalur districts to the MPM to raise its own plantations. In return the industry pays the government 12.5 per cent of the yield. Thus MPM set up its own forestry wing, headed by a Director (Forests) who is a Chief Conservator of Forests. These plantations were raised with technical and financial assistance from the Commonwealth Development Corporation (CDC) and Overseas Development Agencies (ODA) of Britain. While pines, acacia and casuarina are being planted in areas receiving a mean annual rainfall ranging from 900 mm to 3,000 mm, eucalyptus is raised in areas with rainfall less than 900 mm (Nair 1987).

Earlier plantations were raised by the Forest Development Corporations and the yield made over to MPM (see Table 6.2).

In 1980 the GOI brought in the Forest Conservation Act which prohibited diversion of legally constituted reserve forests to outside agencies even though the land had been degraded. It was then that some of the degraded forests were typed as degraded C and D lands and offered to industries. There were no takers. In 1981 the State Forest Act was amended, abrogating the rates entered into agreements with the industries. In the case of eucalyptus the rate of Rs. 24 per tonne with Harihar Polyfibres, was increased to Rs. 205, later to Rs. 264, with retrospective effect. The act also provided for increasing the rates once every two years.

Meanwhile, in 1984-5 KFPC took on the lease of over 87,000 acres of eucalyptus and mixed plantations raised by the FD. The eucalyptus was sold to three allotted industries, viz., HPF, WCPM and MPM for a total sum of Rs. 156.56 lakh. Preference was given to HPF. HPF belongs to Birla. Extraction of eucalyptus would have been much higher but for the writ petition filed by WCPM asking for equitable distribution of raw material in respect of plantations in the Dharwad district. The High Court order directed that the raw material from Dharwad would be shared in the ratio of 52.5:47.5 between HPF and WCPM respectively.

In 1984, when the State Forest Act was further amended, several categories of plantations in which eucalyptus was earlier committed to Harihar Polyfibres were withdrawn in order to meet rural needs.

TABLE 6.2: EUCALYPTUS PLANTATIONS RAISED BY KFDC AND YIELDED TO MPM

Year of plantation	Year of extraction	Area conceded to industry (in ha)	Yield (in mt)
1972	1983-4	1011.33	21,225
1973	1983-4	731.99	15,515
1974	1984-5	710.25	22,171
1975	1984-5	535.64	15,865
1976	1985-6	854.01	20,919
1977	1987-8	1460.16	32,795
1978	1987-8	770.76	9,780
1979	1988-9	221.04	1,904
1980		320.00	3,097
1981		215.60	1,500

Source: KFDC, Shimoga.

TABLE 6.3: EXTENT OF EUCALYPTUS PLANTATIONS RAISED BY KFDC LTD.

Year	Extent (in ha)
1972-9	49,920
1979-85	40,349
1985-6	44,024
1987-8	52,926
1988-90	Stopped

Source: KFDC, Bangalore.

Harihar Polyfibres woke up to the realities (Shyam Sundar and Parmeshwarappa 1987: 192).

The year 1985 also saw Karnataka Pulpwood Ltd. (KPL) being incorporated as a joint sector undertaking with KFPC holding 51 per cent of the shares. There are three representatives from HPF on the Board of Directors. KPL was set up to raise captive plantations for the Birla's Harihar Polyfibres (HPF).

West Coast Paper Mills coped on its own and was rather sore with the privileges bestowed upon the Birlas (WCPM's Forest Manager, private communication, 1985). It had its own plantation of eucalyptus on a degraded hill at Kulahalli about 200 km from its factory. The area is said to have been acquired through a fraudulent deal with its owner (WCPM employee, private communication, December 1985).

KPL had planted 600 ha of eucalyptus and casuarina and proposed to raise another 2,000 ha during 1986. This project was sanctioned by NABARD.

During 1986-7, KFPC changed its name to Karnataka Forest Development Corporation (KFDC). The same year also saw the government increase the lease rent in keeping with order AHFF 9 and FDC 87 dated 10 April 1987, from Rs. 10 per acre per annum to Rs. 25 per acre per annum till 1987, and then to Rs. 187 per acre per annum from 1987 to 1993. The order also required KFDC to take over all FD's plantations. Eucalyptus wood was sold to the three PPI industries, HPF, WCPM and MPM. Expenditure incurred by KFDC till 1987 was Rs. 1,261.5 lakh. Meanwhile KPL entered its second year. According to a government order (FFD 180: 84 dated 15 January 1986), 30,000 ha of land, for raising of eucalyptus and casuarina were to be leased to KPL for HPF, for a period of 40 years. The lease rent was fixed at 12.5 per cent of the produce. KPL was to receive loans from NABARD to the tune of Rs. 700 lakh as well as loans from three commercial banks to cover 10,000 ha of plantations over three years.

Compelled by the ongoing people's movement, KFDC gave away the lops, tops and barks of eucalyptus, free of cost to the villagers. According to KFDC (*Annual Report*, 1984-5: XV) this accounted to 30 per cent of the total yield. Then another 12 per cent was sold to the people through the FD, only 57.5 per cent of the wood going to the industries.

The KFDC *Annual Report* for 1986-7 states that all wastelands, either barren or degraded, were being covered with plantations of eucalyptus, rubber and tea. The cocoa, rubber and tea plantations were merely being maintained, which perhaps means, not expanded. At the instance of the Wasteland Development Board and GOK, KFDC implemented silvi-pastural schemes raising 4,000 ha of wasteland; this was being done for and on behalf of the FD (*Annual Report*, KFDC, 1986-7: VI).

The 1980s saw the GOK launch large-scale social forestry schemes. These are of great significance to this study, even though social forestry was described as forestry for and by the people, with the primary objective of providing the rural population with fuel, fodder, green manure, small timber and bamboo. This comprises community forestry and farm forestry schemes, afforesting common lands, C and D class lands, road sides, canal sides, etc. Seedlings were supplied to farmers free of cost or were subsidized. In 1983 the GOK approached the World Bank (WB) and Overseas Development Agency (ODA) with a proposal for a Rs. 55 crore programme for social forestry. The plan was to raise 29,000 ha of community plantations and 1,20,000 ha of farm forestry through individual participation.

The World Bank suggested planting of 5,000 seedlings per ha of which 46,000 seedlings of *Eucalyptus tereticornis* was for the Social Forestry Programme (Shyam Sundar 1987: 195). Only 2,500 seedlings were planted per ha and the rest were fruit trees in the social forestry programme. This is mentioned only because *Eucalyptus tereticornis* forms a large component of the SF and also because Chandrashekar et al. (1987) hold the Forest Department responsible for feeding the rayon industry, that is Harihar Polyfibres, with eucalyptus. Incidentally, it must be mentioned that *Acacia auriculiformis* is also visible as a prominent road side species and as a part of the Social Forestry programmes both in cities as well as in the countryside in and beyond Karnataka.

Meanwhile, in 1988-9, a public interest litigation was filed against the KPL in the Supreme Court on issues of legality of lease of forest lands by the FD to the company, for growing pulpwood for Harihar Polyfibres. The company officials of WCPM and HPF were irked that people only protested against lands allotted to them and not about the 30,000 ha allotted to MPM which is a public sector unit. Of the promised 30,000 ha, only 3,590 ha were granted to KPL. The FD had also transferred only 36,000 acres of the promised 87,000 acres to KFDC. Till 1988-9 the expenditure incurred by KFDC was Rs. 2,545.62 lakh.

Year	Casuarina and misc. species	Eucalyptus and fodder supply
1986-7	900 ha	1900 ha
1987-8	830 ha	1645 ha
1988-9	915 ha	1885 ha

Source: KFDC.

KPL was closed down in 1991. The total extent of plantations raised by KPL till 1989 is given below:

Out of a total area of 9590 ha permitted by the Supreme Court duly leased to KPL by FD, KPL had raised plantations to the extent of 8,524 ha till March 1990. No new plantations were taken up in 1991 (KFDC, *Annual Report*, 1991: 78).

In the wake of the two government orders (mentioned in earlier sections), raising of new eucalyptus plantations has been stopped since 1990. The available wood from plantations is being sold through traders.

Officials of KFDC reveal that their corporation is suffering from lack of funds and is therefore totally inactive, though for the year 1994 they planned to take up massive bamboo plantation schemes in Uttar Kannada. They also pointed out that it was impossible to raise plantations because local people encroach on the lands and take up agriculture.

Thus, while KFDC is the supplier of eucalyptus wood to HPF, WCPM and MPM, MPM also meets its raw material requirement from its own plantations and HPF from the various farm forestry projects that it has initiated. The Forest Department also supplies the three industries with raw material as represented in Figures 6.1 and 6.2.

TABLE 6.4: DETAILS OF PLANTATIONS RAISED BY THE FOREST DEPARTMENT

Year	Area in ha	
	Eucalyptus	Bamboo
Till 1988	1,25,969	4,892
1988-9	4,949	557
1989-90	Stopped	689

Source: Karnataka Forest Department.

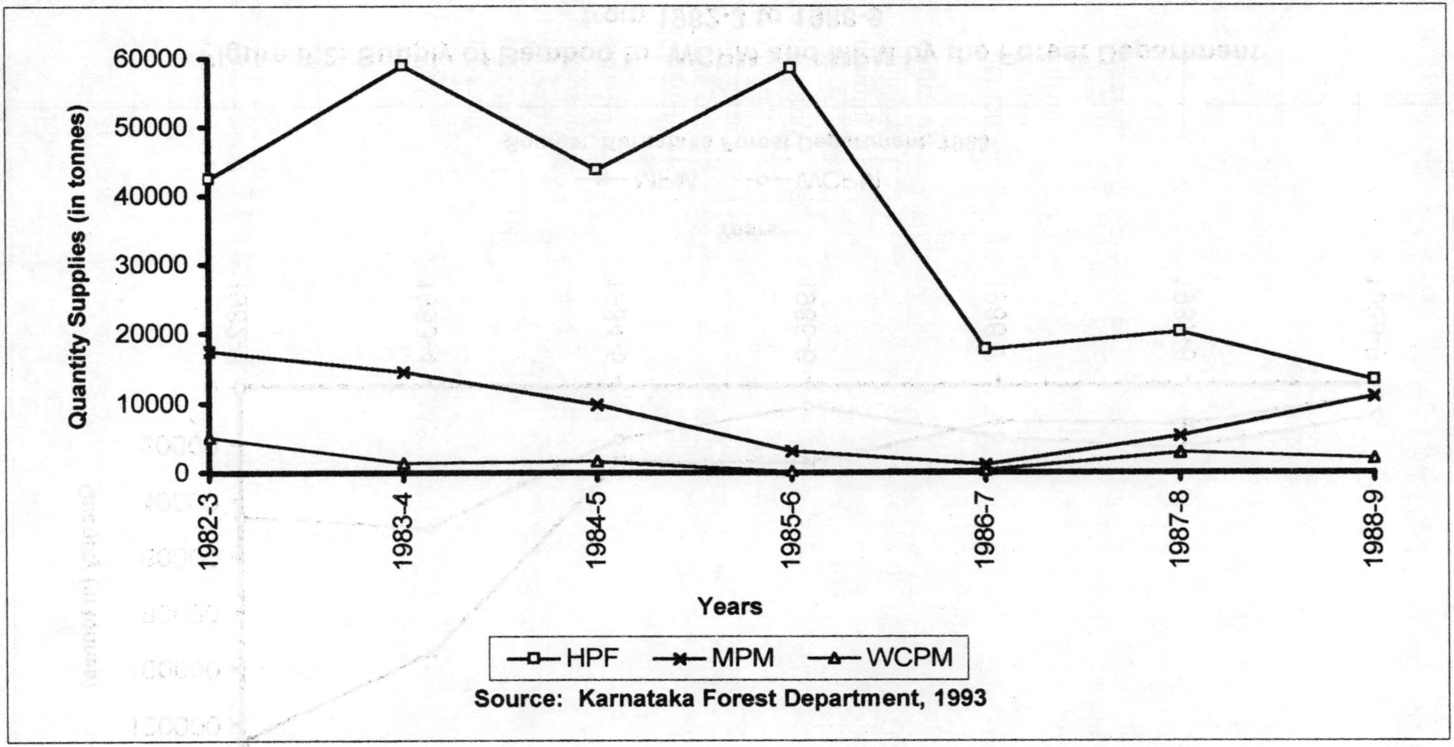

Figure 6.1: Supply of Eucalyptus to HPC, WCPM and MPM by the Forest Department from 1982-3 to 1988-9

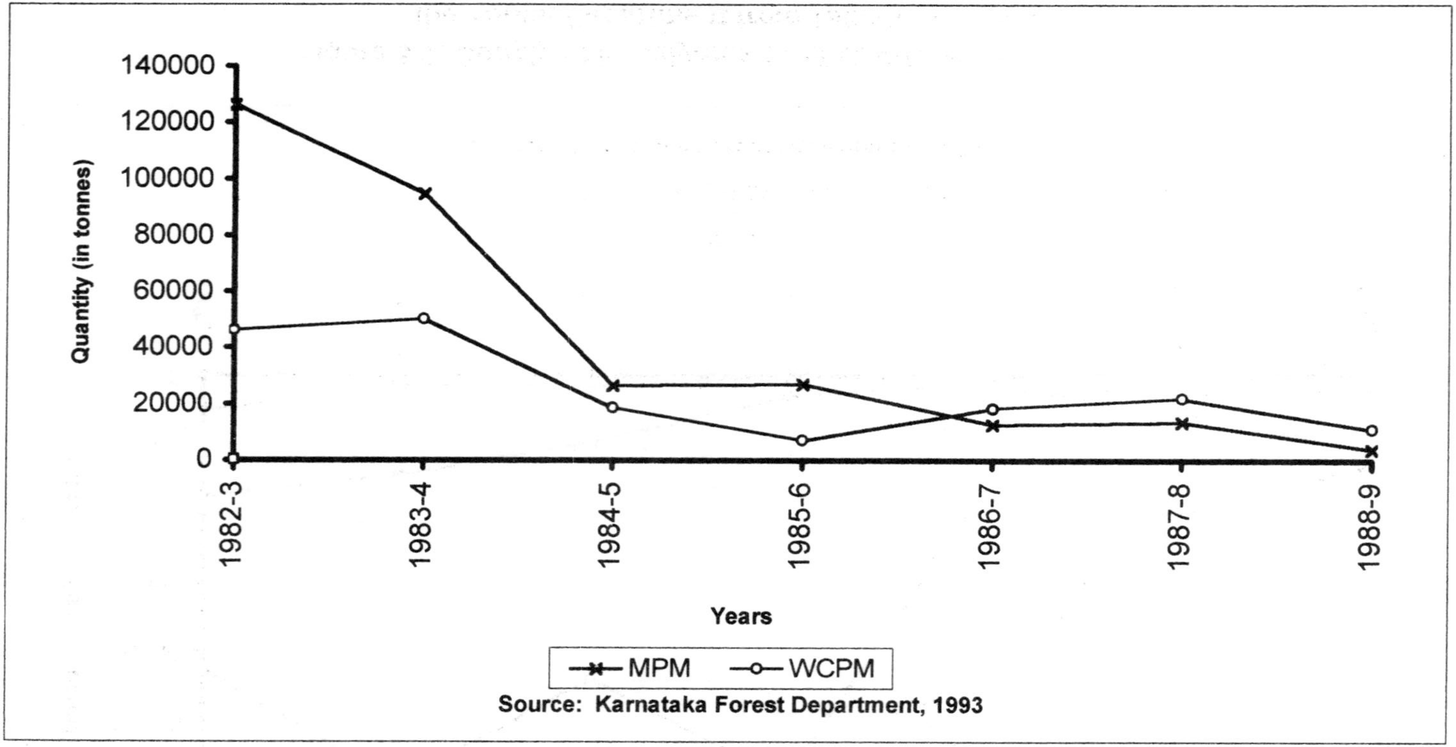

Figure 6.2: Supply of Bamboo to WCPM and MPM by the Forest Department from 1982-3 to 1988-9

SECTION 2 (B)

The Ecological Effects of Plantations on the Forest Ecosystem with Special Reference to Karnataka –An Issue

In this section the question of plantations, especially the ecological consequences of plantations of fast-growing exotic species cultivated to feed the pulp and paper industry, is raised. Large-scale pulpwood plantations were first raised at the recommendation of the FAO specialist who predicted a shortfall of raw material (Von Monroy, 1961) despite the abundance of bamboo that its preliminary investigation had revealed. Incidentally FAO had suggested Coorg in Karnataka as a location for a PPI unit (FAO 1954: 30, 33, 64). This seems to have been counterbalanced by sheer ignorance of the dual system of reproduction and regeneration of the bamboo species, and its rejection as cellulosic raw material as early as 1952 (FAO 1953: 110).

It is not clear why the FAO experts had not cared to read what the British foresters had to say about the resilience of bamboo which could withstand even the trauma of light fires, and give rise to secondary formation, as was noted by Kerala Forester Bourdillon (1892), by Hewetson (1926), and Bergbie (1910), an administrator in the Central Provinces. Stebbing (1921: 400) confessed that 'attempts were also made to kill bamboo by repeated cutting of new shoots'. It is also important to note that despite its resilience, bamboos vanished from almost all the catchment areas of most of the PPI units due to the utter indifference of and, therefore, faulty felling practices by the industry (Gadgil and Prasad 1978). Under strict vigilance of the FD, BILT in Maharashtra and Titaghur in Orissa which have treated the catchment areas with respect, still have abundant bamboo. The team also noted abundant natural regeneration of bamboo in Karnataka after the state government imposed a ban on felling in 1990. Homestead bamboos in Assam and community held bamboos in Nagaland (see Ch. 9) also flourish. These are two important issues which have to be kept in mind while considering the issue of plantations.

The issue of pulpwood plantations on denuded forest lands is illustrated through the experience of Karnataka. There is available information on that region to show that human interaction with the forest is neither new nor necessarily damaging. The nature of damage depends upon various factors. These include the intensity of interaction, the nature of the ecosystem and the species that is introduced. This issue is raised mainly because Karnataka, like the other states in the south, has taken to greening the denuded forests through a variety of plantations including pulpwood plantation of exotic species, as mentioned in the earlier section. Do these species, alien to the local ecosystem, help in regenerating the denuded forests? If they do not, are there any negative consequences? One cannot forget that a ban on felling even as late as 1990, has had some positive effect.

Karnataka has an area of 38,723.6 sq. km under the control of the Forest Department and about 2 million ha that is slightly more than 50 per cent, are in a state of degradation (Forest Department *Annual Report*, 1990-1: 1-4), and this area would be greened by plantations.

Forest Ecosystem

It is necessary to start with the assumption that trees by themselves do not make forests; and that a forest ecosystem is complex. There still exists a largely unknown relationship between the organic and inorganic elements which are delicately balanced to form a protective mantle over a fragile lithosphere, this being particularly true in the tropics. The relative and absolute amount of rainfall as well as periodicity of rainfall strongly influence the ecosystem. The precipitation pattern is a result of atmospheric circulation and topography. Water is an agent of nutrient distribution and loads of it runs downhill. Upland areas are potentially less nutrient rich than the lowlands. Water, after all, is a significant medium of biological activity. Rainfall which runs down in streams becomes a source of ground water. 95 per cent of water is bound up with the earth's lithosphere and in the sedimentary rocks fresh water amounts to only 3 per cent.

One must also look at the relationships in terms of abiotic and biotic. The abiotic backdrop includes the basic inorganic elements, besides physical factors such as gradients, moisture and solar radiation with which the biotic components, i.e. plants, animals and microbes interact. Sunlight is significant for autotropisms of plants and carbon dioxide is assimilated into energy-rich carbon compounds. Chemosynthetic bacteria oxidize simple organic compounds and play an important role in the movement of nutrients in the ecosystem. Under anaerobic conditions they are converted to organic substances which would otherwise be lost through storage in sediments. The heterotropes, bacteria and fungi perform invaluable service for the ecosystem by mineralizing organic matter. The two processes happen concurrently (Kormondy 1991: 41, 50 and 55).

Large-scale disturbance of the lithosphere can set destruction of the ecosystem in motion. But the disturbance of the ecosystem by the people in ancient times was limited in space, intensity and was therefore qualitatively different from what it is today. Homo-sapiens, or to be historically more specific, woman, through her observation and interaction through repeated experimentation with the ecosystem, gained a degree of control over it in order to produce her own food, be it with a hoe or by slash-and-burn, rather than merely foraging and taking what the ecosystem gives. But primitive interaction was limited by the quality of the tools and therefore limited in space; as a result, the negative consequence on the ecosystem was minimal. The forests were vast and there was enough space to move without damaging the same area. Viewed in terms of time, the ecosystem had every opportunity to recover. The people-nature relationship remained harmonious, intentionally or otherwise.

Such a balance was not confined only to primitive forest tribes and their ecosystem but continued even after tools became a part of civilized society. Nineteenth-century District Gazetteers, the colonial documents based on keen observation, written as well as oral, and local history, were meant as guidelines for British administration and control of its most precious colony. The 1883 Gazetteer of North Kanara mentions the prevalence of the ancient tradition of the Haivic Brahmins who carefully tended their areca-nut plantations in the evergreen forest in the hills of North Kanara (M. Gadgil 1978). Gadgil vouches that neither the plantations nor the *Soppina betta*, which provide green manure or mulch for arecanut plantations from the hills around, destroyed the ecosystem. But then no Haivic or for that matter anyone would lop the trees for their *soppu* (leaves) or firewood as and when desired from the *betta*, the hills. Lopping was confined from December to February (A. Gibson 1866: 71), after which the sap rises once again and the trees begin to grow. Forest conservation was a known art preserved through folklore and myth, and so too the rules and regulations which permitted lopping of only the lower branches. Lopping, therefore, amounted in practice to scientific pruning, helping the tree to grow vigorously during the spring and monsoon.

It was not only in the hills but elsewhere too, that people cultivated orchards. Van Somerson was a British Forest Officer in charge of the forests in Mysore state in the late 1860s. He made the first survey and recorded more than 8 lakh trees distributed in 16,000 *topes* (orchards) in 12,138 villages, but noted no destruction of the ecosystem.

Teak Plantation and the Ecosystem

Evidently then, the known major anthropogenic disturbance to the forest ecosystem was created by British colonialists who were desperately in need of durable timber. They had learnt from the Portuguese, the value of teak for shipbuilding and needed enormous quantities for defence and building their infrastructure. Their insatiable greed led to destruction. When teak forests were at the point of exhaustion, Cornelly began large-scale experiments through repeated trial and error methods of regeneration within plantations in Malabar in the early nineteenth century. But Gibson's attempt at Dandeli failed (Stebbing 1923: 563).

Teak is a native, and deciduous in habit. It enriches the soil with its litter and therefore harbours a rich vegetation. Bamboo is often its understorey, and though cultivation of teak *per se* is not rejected, as a plantation crop it certainly is. The process of establishing a plantation is injurious to the fragile tropical hilly ecosystem, as moist deciduous teak forests grow on hill slopes with heavy rainfall, and the loosened soil gets leached out.

Incidentally, the natural teak of Dandeli is regarded as superior to that of Kerala and only second to that of Burma. Brandis, the Inspector General of Forests, was the first to realize the fallacies of replacing devastated forests by plantations. He recommended silviculture as a method of restoring forests to normalcy. Nevertheless, teak was an exception and he wanted it and therefore conceived *taungya*, i.e. interplanting teak with primitive cereal cultivation. This initially failed. It finally succeeded in deciduous forests by the turn of the century through various trial and error methods. 12,000 acres in Mysore state were brought under teak plantations (Stebbing 1923: 565-6).

Other Plantation under the British

Any plantation established on the hillsides damages the ecosystem. Mysore came under direct control of the British for a brief period between 1831 and 1881. The period witnessed the opening up of the upland forests for coffee plantations (Muthanna 1955: 4 40). Even earlier, Stebbing (1926: 22) noted the ecological damage to the Sampangee and Karoor ghats and other places in Wynaad and Annamalai in the neighbouring states due to the deserted coffee estates, a situation which seemed very little better than *Kumri*. When the coffee prices doubled, the coffee planters found it was profitable to take up a tract of primeval moist forest on mountain slopes, getting a bumper crop in the third, fourth and fifth year. But denudation of the soils goes on rapidly. It does not pay 'to cultivate coffee in the same area for many years'. Stebbing asked, 'can we restore the grand old forests with all its climatic influences'? (Stebbing 1926: 22). A thorny wilderness takes its place. Plantations of tea, cardamoms and *cinchona* which also need a similar forest ecosystem were introduced on a massive scale. Such commodities were sold at huge profits in European markets but were controlled by London-based shareholders. Although no studies were available to evaluate the effect of the above listed species *per se* on the ecosystem, yet the high rainfall and steep gradients that are needed for establishing the plantations are always prone to leaching, degradation and damage. Right through 1880, systematic efforts were made to raise plantations of sandal, teak, gallnut and cardamom (Muthanna 1955: 440) also credits Hutchinson for introducing exotics into Mysore in the 1890s particularly from Australia.

The plantations that the British set up were primarily destructive to the ecosystem because of the enormity of the space involved, the location and intensity; secondly because many plantations were single-specied or monocultural. Primarily the plantations were established by clear felling natural forests and firing the area to clear the debris. Burning of slash is tricky. It can provide an initial boost of some nutrients like phosphorous, potassium, calcium and magnesium in the form of ash but nitrogen and sulphur are volatilized and lost. In practice, most of the nutrients released from the ash are lost through leaching before they are used up by the young saplings in the plantation (FAO FD 103, 1992: 124). Heavy tropical monsoons and the gradient contribute to rapid leaching of soil. But this is not the case with the primitive slash-and-burn agriculture. The nutrients are not lost as the fired area is a mere patch in the vast forest. The released nutrients are used up not only by the agricultural crop, but by the surrounding trees in the forests as well as by the stumps left behind after light firing. The stumps are meant to regenerate themselves and they do so and grow to their old glory (Turner 1872, quoted in Francis 1907: 1161-81), or make way for sun-loving *D. strictus* (Stebbing 1921: 36) or whichever be the local species of bamboo.

The other danger to plantations in the nineteenth and early twentieth century were pests, if the species were monocultural or were even aged. *Spike* often threatened disaster and *Hoplaocerambyx spincornis* which are inhabitants of felled logs or moribund trees of *Shorea robusta* were known to have attacked healthy mature trees when that species was extensively established as pure plantations (Gibson: 139-61; Jones, 1977, quoted in FAO 1992: 103). Now FAO (1992: 30), concedes and even explains the reason:

> Natural forests which contain a mixture of species achieve a state of complexity in which vegetation is broadly in balance with pests and disease. An increase in the proportion of any host plant is broadly in balance with pest and diseases. An increase in the level of any host plant in the ecosystem tends to be followed by an increase in the pest or pathogen population, and may reduce the host population.

FAO warns that mono-specific plantations, 'increase the risk of attack by pests and diseases unless other habitat requirements of the attacking organisms can also be managed'. Does the third world have the resources to do so? The warning note on pests is wisdom born of FAO's hindsight, after almost half a century of work on preparing the third world and coaxing it, with enormous loans from various international funding agencies and its own immense research, to abandon natural forests and take to industrial plantation forestry. Can the trend be reversed? We see no evidence of change of heart either in the case of the representatives of powerful PPI or the Indian government and policy-makers, despite the fact that in 1981, FAO discovered that its recommended species did not match with the site.

What brought about a change in FAO's Forestry Wing? Two known factors: the first GOI's 1973 demand for a review, the findings of which were not made public. Added to this was the movement against the Social Forestry Programme with exotic species in the Third World countries. Even more powerful were the movements in the West with its panic syndrome, with its concern for the tropical forests which it regards as the life-support system of the earth as carbon sinks. FAO (1991), now accepts that diversity is an important mechanism against pests and diseases.

This study has inadvertently slipped from the colonial past into the present post-colonial period when forests bruised and battered by British imperialism and further ravaged by Indian PPI and its supportive government, have come under the international persuasive influences of the UN's FAO wing, as well as international lending agencies such as the World Bank, USAID, SIDA, CIDA, DANIDA, IDA and the Commonwealth Development Corporation. All these are actively interested in India's forestry programmes, be it plantations or social forestry for the PPI. Two of the above-mentioned agencies are active in promoting exotic monocultural plantations in Karnataka.

The earth is sending out danger signals, the rural population deprived of its rights to forest produce is restive, environmentalists are protesting and FAO, in the 1992 Forestry Paper 103, has laid down the ground rules for forestry, i.e. plantation forestry. FAO had suspected the dangers as early as 1954 (pp. 61-3).

FAO's Rules on Plantation Forestry 1991

Of prime importance is site fertility. To maintain it:

1. Species must match the site in terms of climate and soil;
2. Site clearance must be appropriate to the species, climate, slope, and rainfall intensity in order to prevent soil erosion;
3. During the establishment phase, the sapling must get a good start with adequate nutrients, moisture and weed control;
4. Protection from pests and diseases.

Ecological Status of Plantations: With Special Reference to MPM Plantations

A detailed account of the species, investment and agents promoting these plantations in Karnataka were provided in the earlier section. Here an assessment of the plantations in relation to their ecological consequences is presented.

Evaluation of pulpwood plantations on degraded forest lands would largely be confined to those of MPM on which the mills' published data is available and also because careful observations of some of the sites in March 1993 were made. This evaluation will be based on the parameters laid down by FAO on what constitutes 'good forest plantations'. The findings of Indian scientists and foresters concerned with forest ecology in order to get a fuller assessment have been used.

MPM is a joint sector industry and therefore its plantation wing is directly under the control of a Chief Conservator of Forests who is assisted by a conservator, and six DFOs or assistant conservators besides forest rangers, field assistants and daily wage labourers. The total degraded forest area allotted to MPM was 30,000 ha. The Overseas Development Agency (ODA) gave technical and financial assistance to the project. This is referred to as the fast-growing species project under captive plantation programme (MPM, p. 1). The plantation activity was divided into two phases. In the first phase 16,666 ha were planted between 1981 and 1989.

The entire area made available to MPM for plantation development was divided in two zones, zone 'A' was a wet zone area with a mean annual rainfall of around 900-3000 mm. A part of the degraded site for plantations are in Malenadu (uplands) on the eastern flank of the Western Ghats. In this zone, tropical pine and bamboo were planted for long fibre, and *Acacia auriculiformis* and *Casuarina equistifolia* for short fibre requirements. It is interesting to note that no eucalyptus was grown following the government policy which prohibited its growth in areas with heavy rainfall (MPM, March 1993).

According to FAO (vol. I, 1980: 1), *Acacia auriculiformis* is indigenous to the island off the north coast of Australia. This species grows fast on poor soils in rainfall of 1500-1800 mm a year and in the dry season, for six months. It adapts itself to a wide range of soils and is successful on steep slopes. It has been introduced successfully in India. Its effect on soils has not been mentioned. The FAO document (1980) on its pulping characteristics states: Its yield is on the low side in sulphate pulping. Unbleached pulp is very dark. It suggests use of additional bleaching which is necessary for cheap writing and printing grade paper.

Casuarina has been discussed in Ch. 5 (Andhra Pradesh), and bamboo in several chapters.

Zone 'B' was a dry zone area with mean rainfall of 900 mm and below. In this zone, *Eucalyptus tereticornis* was planted, but had to be phased out as the yield was low and later was replaced by the more promising *Eucalyptus camaldulensis petford provenance* from Queensland, as a short fibre source.

Confusion and Contradictory Findings on E. tereticornis

Findings of FAO (1980) is reported in Ch. 7 (Kerala) and Ch. 8 (Tamil Nadu) and is therefore not repeated. All that needs to be noted here is that *Eucalypts tereticornis* has exhibited negative reaction to the semi-arid zone and has low resistance to drought FAO (vol. I, 1980: 23).

FAO (vol. II, 1980: 133) found only *E. camaldulensis* has failed to acclimatize in many tropical countries—which includes India. Its yield of sulphate pulp is average but NSSC pulping give high yield.

For nursery studies and experiments see note 2.

MPM reports that its work on the plantation began first with assessing the sites in terms of their gradient soil depth and rocks. The area was then demarcated, and weeds and scrubs were burnt between the months of January and March, as the temperature in this area begins to rise after that (MPM, 1993: 3).

A forester of Karnataka cadre writes,

> Burning not only alters soil moisture and temperature but also affects fertility due to loss of soil-organic matter and has far-reaching ecological effects in soil fauna. Further, the soil gets dessicated and exposed to even greater soil radiation. Whatever soil flora that existed on a partially decomposed leaf litter in the scrub area, and which is generally confined to the top 3.0 cm, is destroyed. Tropical earthworms, an important constituent of soil fauna which can tolerate 20-30 °C, will die with clear-felling and fire, as the temperature in dry gravelled soil is 10-15 °C higher than ambient air temperature. (Yellappa Reddy and Karwari, p. 34)

MPM had the soil at the plantation sites worked by D-50, D-80, D-120 or D-155 bulldozers to open up and loosen hard impermeable top soil to permit infiltration of rain-water. It was meant to conserve moisture and root penetration. Steep slopes where bulldozers could not be used were manually worked. Manual labour was also engaged to crush the clod and to form 4-metre long trenches.

With the onset of the south-west monsoon, four to six month-old saplings were planted in small pits scooped out along the rip line at a spacement of 1.5 m for acacia and 2 m for pines in the wet zones. Eucalyptus was planted in the dry zones and the spacement was 1.5 m. The area was weeded during the monsoon and inorganic fertilizer NPK 15:15:15 at 25 gm per plant and rock phosphate at 50 gm per plant was applied. The earth around the plant was loosened to conserve soil moisture.

In November and December, towards the end of the rainy season, soil was dug on both sides of the rip line to absorb and conserve moisture from the occasional rain. In January the area was again weeded and fire lines laid. Aftercare operations in the second and third year were weeding, digging and cutting fire lines. Fertilizers NKP and rock phosphate (15:15:15) at 35 gm per plant were applied again only in the second year. No more fertilizers were used. Acacia and eucalyptus raised in the first phase were harvested in 1991. The yield per ha of acacia in the wet zone when air dried and debarked was 79 tonnes per ha while *E. tereticornis* yielded only 12-18 tonnes. Eucalyptus, of course, was grown in the dry zone.

Till 1989, the area under acacia was slightly higher than *Eucalyptus tereticornis* namely 7,392 ha and 6,549 ha respectively, and *E. camaldulensis* on 521 ha. From 1991 onwards to the projected year 1993, acacia will clearly exceed *E. camaldulensis*. These are 6,794 ha and 2,050 ha respectively. No new eucalyptus plantations have been planted but as many as 589 ha have been earmarked for tropical pines (MPM, p. 5, refer to Table 1, earlier section).

Observations

MPM's *Acacia auriculiformis* plantations in the wet zone in Malenadu appeared to be in good shape and were well-tended. The new broad leaf variety *A. mangia* is still in an experimental stage but those planted out were also doing very well. The plantations were fed with inorganic fertilizers in the first two years. What was particularly noticeable was the thick pile of leaf litter in both varieties. This does not tally with the observation made by A.N. Yellappa Reddy and G.V. Sagar elsewhere. They found *auriculiformis* plantation in the high 2,000-2,500 mm rainfall area and also in degraded forests in Kegari. Uttar Kannada had a high litter fall and degraded fast due to the activity of bacteria *Azatofactor* and *Actinocytes*. They found no fungal activity and ascribe this to the increased pH (7.2) in the plantation. K.V. Sankaran et al. (1992) came to different conclusions in their still unpublished project KFRI 126/89 on '*Litter Dynamics*, Microbial Association and Soil Studies in *Acacia auriculiformis* Plantation in Kerala', also a high rainfall area.

Litter decomposition is an important component in the nutrient cycling process in a forest ecosystem. The rates and pathways of litter decomposition are determined by the qualitative and quantitative composition of the decomposer community, physical environment and the quality of the resources that animals and micro-organisms utilize (Swift, Heal and Anderson 1979; Sankaran 1993: 226). But MPM records indicate that the degraded forest land on which this plantation was established was cleared, burnt and levelled which would have given little chance for the survival of micro-organism so vital for the decay of the litter. It is possible that due to heavy rains and high altitude the scrub cover had earthworms and tropical termites to maintain the porosity of soil.

K.V. Sankaran et al. (1992) also found that the decay of litter was faster in fertile sites than in the degraded ones. But it was everywhere much slower as compared with the litter of teak, *Xylia xylacarpa*,

Albizzia falcatoria and *Terminalia peniculata*. In order to ascertain the effect of acacia plantation on soil properties, they made comparative soil analysis. They found no significant difference between acacia plantations and adjacent fallow lands in forms of physical and chemical properties of the soil and nutrient content, except for the soil pH. The average pH of acacia plantation was 4.7 and fallow 0.5. They also found that acacia forms V.A. Mycorhizol association in Kerala soil. The high level of colonization of acacia roots by VAM fungi shows that the plant is dependent for its establishment and growth and corresponds with the silt content, clay content, porosity and bulk density of soil. *Glomus spp.* was the dominant VAM fungi.

There is a tendency to deny degradation of soil in acacia plantations on the ground that acacia roots form nodules typical of nitrogen-fixing flora. But nitrogen-fixing agents use up the nitrogen they manufacture and only release it through the leaf litter. For nitrogen-fixing agents to make positive contributions even to stand growth, the site conditions must be suitable, the soil has to be deficient not only in nitrogen but also in other nutrients especially phosphates and moisture (Sprent 1985, quoted in FAO 1992: 12).

Nitrogen is derived from mineralization of organic matter by fungi and bacteria in a three-stage process, first to NH_3, then to (NO_2) and finally (NO_3). Nitrogen is introduced into the nutrient cycle by working of nitrogen-fixing bacteria. If the nutrients are held in the litter and are unavailable to the top soil solution then nutrient deficiency will occur. The team's observation showed there was no sign of decay of the thick pile of leaf litter. There was no undergrowth, not even weeds.

The Case of Eucalyptus Plantations

The first recorded evidence of eucalyptus plantations in the country dates back to 1832, *E. globulus*. Planted in the Nilgiris, it was meant for fuelwood to keep the British troops warm. Several species of eucalyptus were attempted to be introduced at various points of time during the nineteenth and early twentieth centuries in different parts of the country with varied degrees of success. By the mid-twentieth century, the eucalyptus hybrid proved to be the most common and responded well as an odd ornamental tree in gardens in widely different conditions of soil and climate. By 1962 this hardy species gradually replacing *E. citridora* and *E. camaldulence*, and *E. hybrid* or *tereticornis*, was widely introduced into forests to provide raw materials for PPI.

That year, many parts of the moist deciduous and even evergreen forests of Kerala were converted to *E. grandis* and *E. tereticornis*. Elsewhere, for instance, in Madhya Pradesh only in the inferior forests, and in Maharashtra in barren and degraded forests, was *E. hybrid* planted. In Uttar Pradesh large-scale plantation of *E. hybrid* was started in the Terai and Bhabar areas containing open and miscellaneous species (Chaturvedi 1976). Everywhere it was *E. hybrid*, thanks to the World Bank. By 1969 about 25,40,000 ha had been planted with this species and further 80,000 ha more per year were contemplated (Chaturvedi 1976: 57).

E. tereticornis took up so much space in Karnataka in forests and elsewhere with no insights into the species-site relationship that it provoked an interesting controversy. Monocultures of eucalyptus were being promoted with no insights into the site-specificity of the species. The credit of first raising this issue in India—mining of water by eucalyptus species goes to Vandana Shiva and Bandhopadhyaya in 1981 and 1985, followed by others like D.M. Chandrashekar.

FAO was provoked. If in 1969 FAO had put down the failure of eucalyptus plantations to lack of trained personnel and obsolescence of administrative machinery (FAO 1993: 114), in 1985 it marshalled professors and consultants from Spain, Brazil and France for advice, from UK and France for working out a bibliography and SIDA to carry out a study on the environmental impact of eucalyptus species.

The FP Report, 59, 1985 opens angrily, weighs the species in terms of its relations to the water cycle, its capacity to stop or not stop erosion, its effects on different soils and their nutrients and the nature of its competitive character. It arrives almost but not quite dispassionately at interesting conclusions.

On the question of water consumption and the recharge of aquifers it states: (a) Lateral spreading and depth of penetration of root systems of the eucalyptus vary with species and this has to do with intensity of water uptake. (b) In regions of deeper soil, water deficit created by eucalyptus forests seems to be in the region of 250 mm/year. In comparison with crop or pasture land this means eucalyptus forested lands would yield approximately 70 mm/year less stream flow or ground water recharge. (c) The effects of soil moisture reserve of eucalyptus plantations apparently start to appear at the age of approximately 4 to 6 years. (d) Transpiration rates differ among eucalyptus species varying between 20 litre to 40 litre per tree/day. (e) Some eucalyptus species have not developed mechanisms for controlling higher rates of transpiration (FAO 1955: 19) and most eucalyptus are not good trees for erosion control (FAO 1985: 21).

It is important to note cropping of eucalyptus on short rotation, especially if the whole biomass is taken, leads to rapid depletion of nutrients in the soil (FAO 1985: 55). The effect of uncropped eucalyptus on soils depends on the state of soil in which they are planted. There is evidence that eucalypt litter decomposes less well because of less activity of soil microbes and termites (p. 54). Eucalypt plantations have less diverse flora and fauna than indigenous forest (p. 50). The Report (1985: 52), also points out the disadvantages: (a) unsuitable for browsing for fodder, (b) can displace agriculture from land suitable for agriculture (Punjab, Haryana, west Uttar Pradesh—the green belt is a case instance), land seizure for planting and displacement of labour, (c) can lead to fuelwood becoming a market commodity where previously it was free goods. In other words FAO's assessment tallies with that of the team. It may be added that coarse grain that is cultivated on impoverished soil may cease to yield if eucalyptus plantation are raised. We also agree with FAO's (1985: 55) conclusions that 'eucalyptus should not be planted, especially on a large scale, without careful and intelligent assessment of the social and economic consequences . . . sympathetic examination of the ecological circumstances and the need of the local people'.

The utility of eucalyptus for making sturdy poles, industrial fuelwood with high calorific value, medicinal leaves and cellulosic content for paper is not denied. But its advantages have to be weighed against its relations with the ecosystem. It has now ceased to be a controversy. Most foresters now openly concede that eucalyptus was a folly. There are exceptions of course. Shyam Sundar (1987), for instance, holds that eucalyptus roots do not penetrate deep and therefore do not pose a threat to the water table. He was impressed by the growth of eucalyptus in the Brazilian plantation. FAO reports that it surpasses its growth in its native Australia. It is also reported to grow well in Chile and in several African States that fall in the southern hemisphere. The head of the forestry wing of FAO agreed with the team's observations and noted that the growth of eucalyptus in Portugal and Spain is not too satisfactory (Lanly, personal communication, October 1995). There is no theoretical explanation for this strange phenomenon, it is an empirical fact. Splendid old specimens of *E. globulus* in Nilgiris and *E. citridora* in the region can be found. The Nilgiris and Annamalai are only 11° and 10° North of the equator.

Thirawat (1956: 661), one of the participants in the FAO team to Australia in 1952, gives an impression that extensive plantations started in Brazil, Union of S. Africa, Madagascar, 'Belgium Congo', etc., in that order, were because all of them belong to the southern hemisphere and lie in the same range and latitude as Australia, and the second biggest group of countries utilizing Eucalyptus are the Mediterranean climatic group with winter rainfall and sclerophyllous forests. Thirawat is sceptical about its success in Asia, but notes that it grows well in cool Nilgiris, though he has no explanation why it does. *E. globulus* grows in temperate regions in Australia.

In Karnataka, bamboo grew naturally and eucalyptus was cultivated in forests as industrial raw material for PPI till 1978. Cultivation of eucalyptus was because of strong pressures from the central government which came in the form of indicative plans and resources made available for the purpose, sometimes to the total exclusion of forest development programmes. PPI's demand for pulpwood was insatiable. The Third Plan recommendation was based on the FAO consultant J.A. Von Monroy's suggestion, that 0.6 ha or 1 per cent of the most productive forest zones should be selected for a ten-year programme of plantation with fast-growing species, at the annual rate of 60,000 ha for intensive management so as to ensure 4.5 million tonne of uniform industrial raw materials from 1975 onwards. This was to change the pattern of raw material consumption of PPI from predominantly bamboo to predominantly hardwood with long fibre component such as eucalyptus.

Large plantations to meet the demands for industry, fuelwood and other forest products were undertaken with 100 per cent central assistance to the state governments to encourage them to grow quick-growing species; the result being that 87,000 ha of exotics were raised in Indian Reserve Forests.

The fourth plan prescribed linking up of forest development with forest-based industries to achieve self-sufficiency especially for PPI wood panel products, matches, etc.

The fifth plan aimed at a 'dynamic programme' of production forestry. This was to be achieved by clear felling larger areas of forests and creating forests with institutional financing. Created forests were to cover 2 per cent of the forest area by first clear felling it. The last clause was recommended by the National Commission of Agriculture in 1976, and was to be achieved by setting up one or more Forest Development Corporations which would qualify for institutional borrowing. The Karnataka Forest Department set up four such corporations—details have been discussed in the earlier section. What is of relevance in this section are pulpwood plantations, and the impact of the species, especially of *E. tereticornis*, on the ecosystem.

Jose Kallarackhal, in his still unpublished findings of water-use, photosynthesis and growth of the two species of eucalyptus, *E. tereticornis* and *E. grandis*, shows that these species in all the localities in Kerala indicate minimal water stress through the year. This he finds is reliable evidence of deep extraction of water through their roots. In other words, he confirms the fears expressed by Shiva (1981). Kallarackhal found that spacing of trees is important. When planted at 2,500 stems per ha, *E. tereticornis* transferred 55 per cent of the water coming in as rainfall. But it transferred only 34 per cent of the rainfall when planted at 1,090 stems per ha. This question is raised here, as it is sometimes argued that some tree cover is better than no tree cover at all, and in degraded land the hard and adaptable *Eucalyptus tereticornis* is the only answer, barring perhaps *Acacia auriculiformis* today.

Is this a correct answer to degradation, or would such species which mine water and survive, further degrade already degraded land? FAO, on the basis of its evaluation of eucalyptus plantations and other such fast-growing trees in the entire third world, now fears that monospecific plantations carry the risk of catastrophic destruction in the from of soil degradation, apart from being attacked by pests and diseases. It also finds that although investments have been large, the achievements have been small in comparison to the predicted demand for industrial pulpwood. Second rotations are common and on some sites even third rotations have been established but the yields have tended to decline. Why? Are these not indicators of depletion of soil and water?

With the accelerated destruction of natural forests there is a fear that forest plantations may contribute to the process of environmental degradation (FAO 1992: 1-3) *E. tereticornis* is even more disastrous than *E. grandis*; it has no stomatal control like the latter in response to atmospheric dryness (Kallarackhal 1992).

Studies carried out by K.V. Sankaran to get a comparative estimate of decomposition of leaf litter of *E. grandis, E. tereticornis, T. grandis* and *P. falcatoria. Albizzia* in Kerala show that teak litter decomposed rapidly as compared to others. Decomposition of eucalyptus litter was the slowest. Leaf litter from eucalyptus differed statistically from *Albizzia* and teak with respect to bacteria per gram of litter. The number of fungi per gram of litter differed significantly between eucalyptus and *Albizzia*, and that of acatinomycetes between teak and eucalyptus. This question of litter decomposition has been raised, because it is an important process of soil formation and nutrient cycling. The rate and pathways of litter decomposition is determined by the qualitative status and composition of the decomposer communities, their physical environment and the quality of the resources that animals and microbes utilize. It is possible that the slow rate of decomposition of eucalyptus litter is due to the presence of polyphenols and volatile terpenes in the leaves which are known to inhibit microbial enzyme action. The hard texture and waxy coating of eucalyptus leaves are additional factors which inhibit decay (Sankaran 1993: 225-42).

There is still another negative factor, viz., the allelopathic properties of eucalyptus. Compared to indigenous trees the plant diversity, biomass and density of growth under *E. tereticornis* is poor and therefore should not be recommended for farm forestry. Even as field-bund-plantation eucalyptus affects many varieties of crops (Kohli 1990).

A.N. Chaturvedi (1976: 58), reports three damaging fungi in eucalyptus plantations *Ganoderma lucidum*, *Cylindrocaladium quinquesptum* and *Corticum salonicolor*. The most effective control against pest is diversity [Perry and Muqhabe 1989 (26): 123-40 quoted in FAO 1992: 31]. No studies are available on the effects of tropical pines on the forest ecosystem. *Acacia auriculiformis* and eucalyptus are not despised because they are exotic. The latter which has survived and established itself for nearly two centuries cannot be called exotic. More recent studies on their water use and leaf litter have shown that as monospecific plantations, they carry greater risks particularly in degraded forest land.

Yellappa Reddy has suggested that agaves are an inexhaustible source of cellulosic raw material for paper (personal communication, 1993). These grow even along the degraded train tracks and can be cultivated. They are imported and cultivated as a commercial crop in East Africa (Lock 1962). There is no available material on their water use pattern. But they yield strong paper, says FAO expert (personal communication, October 1995). Elsewhere, the same author says that *D. strictus* produces higher biomass at lower planting densities (500 per ha under better sites and management than *A. auriculiformis* and *Casuarina equistifolia*, which are planted on degraded open sites at high planting densities (5,000 plants per ha) in the high rainfall areas on the western ghats in Uttar Kannada. Bamboos are eco-friendly local species. They could be almost inexhaustible and also they grow as an understorey. Bamboo should be encouraged to grow in a mixed forest to maintain safe biodiversity and if harvested with care would provide the necessary wood for paper. They have other multiple uses. If sustainable development is significant, then at the present state of knowledge the local species of bamboo which match with the site condition is the answer provided, if tended and harvested with care.

SECTION 3 (A)

Mysore Paper Industry (MPM)

The meandering Bhadra river flowed lazily ankle deep; lush green fields seen on either bank in the serene countryside, the mute witness to the changing panorama of Bhadravati, called Paper Town in 1935. Its surrounding forests abounded in bamboos.

The MPM was set up as a limited company with a capital of Rs. 25 lakh in 1937. Ten per cent of the capital was subscribed by the Government of Mysore and the rest by small shareholders. The government also extended various other concessions and facilities such as free land for construction of the factory, bamboos supplied at concessionsal rates, delivery of bamboo at the mill site, and also provided water and power. The government agreed to purchase paper from MPM to the extent of its requirements, subject to the quality and price being satisfactory. The plant is located at Bhadravati and is designed to produce about 5,000 tpa of typewriting, writing, printing, *badami* and imitation art kraft paper, etc. Besides making its own bamboo pulp, a rag plant for the manufacture of rag pulp, and a soda recovery plant were also set up. Its requirement of chlorine was also prepared at the site through electrolysis of common salt. In other words, the paper plant was made self-sufficient. MPM is the first mill to set up such a continuous chain of backward linkages.

To begin with, the Mysore state held 10 per cent of the equity and was closely involved in the management of the company. It had the authority to nominate three of the nine members of the Board including the Chairperson and the Managing Director. The capital structure of MPM from 1961-2 onward is given below.

CAPITAL STRUCTURE OF MPM

In 1961-2, 10,000,000 Right equity shares were issued at a premium of Rs. 2 per share; in the proportion 1:4.

In 1963-70, 3,116 shares were forfeited.

In 1975-6, 2,66,666 shares were allotted to KG on conversion of a loan.

In 1979-80, 3,116, forfeited shares were reissued to KG. 15,00,000 right shares were then issued at par in the proportion 1:1. Only, 9,25,081 shares were taken up. The balance of 5,74,919 shares along with 42,21,900 shares were allotted. 9,00,000 bonus shares were issued in the proportion of 3:5 to shareholders before the rights issue.

In 1980-1, 32,78,100 shares issued to NRIs or joint stock companies owned by them by private placement with repatriation rights, 12,14,000 shares allotted to NRIs by private placement.

In 1982-3, 26,64,100 rights equity shares were issued at par in proportion 1:4 which include 20,64,100 equity shares earlier issued to non-residents of Indian origin but were not taken up. Only 20,12,270 shares were taken up. Out of the remaining 6,51,830 equity issue, 5,00,000 shares were allotted to KG.

In 1983-4, the remaining 1,51,830 shares of the 1981-2 issue were allotted to KG, 5,03,510 shares were allotted to farmers and KG in February 1984. 2,78,50,000 shares were allotted to KG in March 1984, part of which was on conversion of loans, the remaining being the unsubscribed portion of the shares issued for the sugar project. 1,92,61,966 shares were allotted to financial institutions on conversion of accrued interest and were due to them up to 31 December 1983.

In 1984-5, 1,520 shares were allotted to New India Assurance Co. Ltd., as on 15 March 1985, on conversion of interest accrued and due to them up to 31 December 1982.

In 1985-6, 20,95,431 equity shares were issued at par.

In 1986-7, 6,11,185 equity shares were issued at par to KG on part conversion of loans.

In 1987-8, 9,56,520 equity shares were issued at par to KG on part conversion of loans.

In 1989-90, 10,00,000 equity shares were reissued at par to KG.

In 1990-1, 2,50,000 shares were issued at par.

Source: SEOD, Bombay.

After the Second World War the capacity of the mill was raised in stages to its present levels of 26,000 tonnes a year, but the big change in the character of the company came in 1975 when it was invited by the Government of India (GOI), to consider starting manufacture of newsprint. A few years earlier a private sector group (Jalan Brothers) had acquired a substantial shareholding in the company and the proposed newsprint venture became the subject of disagreement between this group which opposed it and the other shareholders. The private group was outvoted and its shares were subsequently acquired by the state government, which now holds 56.2 per cent of the equity. The rest of the shares were spread widely among small investors, who have no say in the management of the mill. There are interesting and sharp differences in its functioning, its method of modernization, its product mix, and even the method of generating its raw material from those of the private sector.

Operation of MPM

Mysore Paper Mills started its operations in 1937 with a capacity of 400 tonnes a year of printing and writing papers. It was regarded as an example of an 'enlightened' approach to industrial planning which was pioneered by the Maharaja of Mysore in the early years of the century. It provided local employment and made 'good' use of the forest resources, mainly bamboo, of the Shimoga region.

There is no information for the period from 1938 to 1947 except that production started with two paper machines. In 1948-50, Mysore Paper doubled its annual production capacity from 4,000 to 8,000 tonne at a cost of Rs. 80 lakh. The second expansion costing Rs. 5 crore was carried out in 1964-6, when a third machine was introduced. The annual production thereby went up to 18,000 tonnes. In the very first year after expansion, the annual production shot up to 20,000 tonnes (Ananthanarayan 1979).

Meanwhile, in 1962 the price of paper was revised. It was raised to Rs. 60 per tonne. In 1965-6 production increased to 13,500 tonnes. Sales, however, were not brisk for some months of the year, leading to accumulation of stocks and resulting in a strain on the liquid resources of the company. The profits earned during the year were small due to a further increase in the cost of production, without a corresponding increase in the selling prices, as stated in the company's *Stock Exchange Official Directory* (*SEOD*).

The company's production in 1966-7 amounted to 20,000 tonnes, exceeding the annual rated capacity of 18,000 tonnes. The entire requirement of pulp was met from the company's own production, thereby eliminating purchases of indigenous and imported pulp.

In 1967-8 production amounted to 20,427 tonnes. The cost of production continued to rise during the year without there being any improvement in the prices of paper with effect from 3 May 1968, in response to repeated demands for a revision of the selling prices of paper (*SEOD*). Consequent upon decontrol, prices of paper increased with effect from 13 May 1968. With the lifting of price control, all the private sector units without exception introduced new machines, or renovated their old machines to produce value-added speciality paper, forgetting their promise to supply the allotted quota of cultural paper. Sheshayee and MPM were the rare exceptions. They did not change their product line.

MPM's production increased continuously in 1969-70 and 1971-2, which the company attributes to favourable selling prices. In 1972-3, production declined to 22,973 (from 24,076) tonnes, perhaps only due to the 25 per cent power cut imposed from October 1972. The power problem further led to a fall in production by 596 tonnes (22,377 tonnes) in the following year 1973-4. There was a parallel decline in the private sector which had ceased to produce cultural paper, thus creating an intense paper famine. At this point, viz. 1974-5, MPM's production rose by 16 per cent and reached a record figure, *SEOD* (1983) interprets this as due to favourable selling prices. MPM's enhanced production was not sufficient to meet India's growing demand. Production of paper was under an almost total monopoly of big paper units which had ceased to produce cultural paper. The Government of India had no alternative but to bring in the Paper Control 1974 and 1975 Acts.

This was a politically unstable period with the Emergency being declared by the government. The big PPI units responded with alacrity and overproduced cultural paper creating an unexpected slump. Due to this unexpected slump in the paper market, MPM's production declined in 1975-6, resulting also in heavy accumulation of stocks. MPM shut down its plant during the year for purposes of annual maintenance and installation of new plant and equipment under the optimization scheme.

In spite of sluggish market conditions, the high incidence of overproduction by PPI creating an unexpected slump, and the excise duty and stringent money market conditions, the company sales of 25,939 tonne of paper in year 1976-7 actually resulted in an improvement in its profits.

In 1976 MPM began to think in terms of a new project costing Rs. 85 crore. It envisaged manufacture of 75,000 tonnes of newsprint of which 15,000 tonnes would be magazine grade. The project was expected to be commissioned by October 1979.

The newsprint project is also believed to have been initiated at the instance of the union government. MPM opted for the newsprint plant in place of its earlier one to increase its 24,000 tonnes paper capacity to 70,000 tonnes. At this point of time the project was modified to provide for an addition of a 2,500 tpd sugar plant as part of the paper complex in order to generate raw material.

However, the newsprint plant was a brand new one and not an expansion or modification of the old paper mill. The paper mill stands at a little distance from the newsprint one. A feature of the newsprint project was the employment of the Chemi-thermo-mechanical method said to be one of the most modern for processing eucalyptus and mixed hardwoods available locally for pulp making. This process, in contrast to the private sector's wasteful sulphate process utilizes as much as 88-90 per cent of cellulosic raw material. It is particularly useful for manufacture of newsprint. Incidentally MPM was expected to produce 1,12,500 tonnes, 75,000 tonnes of newsprint and 37,500 tonnes of paper. MPM has also modernized its bleaching technique from the polluting chlorine to a more advanced hydrogen peroxide. The effluents from the new pulping process have a relatively smaller quantum of chemicals than the old sulphate process.

Newsprint is run on a single large paper machine. It is still not totally automated but will soon become so. We noticed neatness and cleanliness which was in strong contrast to Tamil Nadu Newsprint Limited. There is a general feeling among the other units that TNPL is way above others in production techniques. This was not our observation (see report on TNPL). We say this also because the G.M. of MPM maintains that, 'TNPL competes with us on the basis of one latest machine used in the production of paper as well as newsprint requiring less labour and less power resulting in high profits' (personal communication, February 1992).

Setting up of the new plant was not without its problems. To begin with, the cost was estimated as Rs. 100 crore financed by term loans of Rs. 32 crore, term loans from Chemical Bank, New York, of Rs. 43 crore; loan from the state government of Rs. 5 crore; additional share capital of Rs. 6 crore and internal accruals of Rs. 14 crore. It was expected that the newsprint plant at Bhadravati would save over Rs. 32 crore of foreign exchange annually (*Economic Times*, 1978). Mr. Saifullah, the Chairperson of MPM, had claimed that the newsprint would be of international standards, and the price slightly higher.

The newsprint project took longer than expected. It went on stream in 1981, three years and nine months after the receipt of the industrial license. The result of this delay was an increase in the cost of the project to Rs. 150 crore from the initial expected cost of Rs. 100 crore. The mill has a single machine, the largest paper machine in India which started forming a continuous web of paper at a time when the newsprint situation in the country continued to be bleak with the demand reaching 3,50,000 tpa, while MPM's installed capacity remained only at 60,000 tpa. Trial production was carried out with imported pulp from Canada because of the delay in setting up its own pulping plant. A new bamboo pulp plant required another Rs. 21 crore. The advanced and eco-friendly pulping technique used by this mill has already been discussed. MPM approached the IDBI in 1981 to finance the expansion scheme. This sparked off a new row between the two, with the IDBI being none too keen to invest, especially after the company cost overrun pushed up the project cost from Rs. 100 crore to 150 crore. The MPM had to argue strongly for setting up its own pulping plant. It contended that if the scheme was not funded, it would place the company permanently at the mercy of imported pulp (*Business Standard*, 1981).

In order to achieve the licensed production of 1,12,500 tonnes annually (75,000 tpa newsprint and 37,500 tpa other papers), the two plants, the old paper and the newsprint required 1,26,000 tonnes of bamboo; 88,000 tonnes of lops and tops of hardwood per year. MPM hoped to meet a major portion of this demand through silviculture-based operations in bamboo concessional areas and the raising of captive plantations in its leased areas. Towards this purpose, MPM had taken on a lease from the Government of Karnataka 75,000 acres of land (30,000 ha). The company had plans to take up another 14,000 ha in a period of five years commencing from 1982.

Production of newsprint started in 1981-2. In 1982-3, production of writing and printing paper was permitted to decline to 17,305 tonnes from 23,934 tonne, while that of newsprint was raised to 34,000 tonnes from 6,770 tonne. In 1983-4 production of both writing and printing paper and newsprint increased to 21,560 tonne and 53,907 tonnes respectively. The cultural paper division and the newsprint division utilized 76.34 per cent and 95.22 per cent of their installed capacities respectively. But the company claimed that but for power cuts ranging from 25 per cent to 60 per cent, the performance could have been still better (*SEOD*). During the sugar crushing season the power consumption for paper goes down. MPM needs this bagasse for production of paper.

The company's annual report for the year 1985-6 states that the total machine production of paper and newsprint at 1,06,978 tonnes as also the total finished production at 99,903 tonnes was the highest ever attained.

The Production and Price Control Acts of 1974-5 were lifted in 1986 after a long period of shadow-boxing between the private sector PPI and the Government of India. PPI had been modernized in the last phase in anticipation of lifting of controls and was ready to take off, but the accumulated stock of small-scale industries had filled in the need and the Government of India's policy had created a glut in paper.

The company was unhappy that it could not maintain its progressive trend of higher output in the year 1987-8, mainly because of the depression in market conditions for cultural paper; this was coupled with a cost-push caused through steep rises in the cost of raw materials, chemicals, and power, the main elements of its total input. Because of these reasons the total output of paper and newsprint was 94,730 tonnes (*Annual Report*, 1986-7). The newsprint production fell because the Government of India again chose to favour the private sector. It helped the private sector to get rid of its accumulated stocks by permitting sale of cream wove paper to the newspaper industry in 1988 (see Ch. 3, NEPA for details).

The total production of cultural paper and newsprint in the year 1988-9 was 1,02,914 tonnes, up from 94,703 tonnes in 1987-8. The newsprint production at 84,855 tonnes was the highest achieved. The company stated that production of cultural paper was remunerative only if the use of bagasse pulp was over 75 per cent. Because of the trial runs of utilizing maximum bagasse pulp in the cultural paper mill, the paper machines had to run at a lower speed which contributed to lower production. Action was being taken for revamping the cultural paper mill to make it suitable to use a higher percentage of bagasse pulp, as also to increase the output of bagasse pulp for optimizing production of cultural paper (*Annual Report*, 1988-9).

During the years 1989-90 and 1990-91, while there was a decline in the total production of cultural paper and newsprint, a significant highlight of these years was improvement achieved in the use of indigenous pulp for newsprint manufacture. MPM thereby reduced its dependence on expensive imported pulp and saved precious foreign exchange. It also reduced the grammage of newsprint to 49 gsm for the first time. Use of imported pulp came down from 27 per cent in 1989-90 to 17 per cent in 1990-91 (*Annual Reports*, 1989/90-1990/1).

A significant improvement in the company's performance with

higher production of newsprint (82,440 mt) in 1991-2, was attributed to the higher usage of indigenous pulp from its captive plantations (*Annual Report,* 1991-2). The production pattern of MPM from 1966 onwards is presented in Figure 6.3.

MPM has started using more acacia instead of eucalyptus for cold soda refiner mechanical pulp, mainly due to the success achieved through its R and D efforts which established that acacia gives better strength characteristics to the pulp thereby improving the machines' runnability. The brightness of unbleached pulp from acacia is said to be much higher resulting in saving chemicals for bleaching. Use of 63 per cent acacia and 37 per cent eucalyptus made yields comparable with 100 per cent eucalyptus. Apart from this, there is great scope for MPM to further increase production by reducing the downtime[2] which is the highest for MPM, as a result of which, as many as 87-90 days are lost each year. The reason for this unusually high downtime at MPM is the shortage of raw materials, power failures and utilities problems, mainly due to inadequate availability of coal (*Journal of Industry and Commerce*, May 1993).

Newsprint Control Order and Price of MPM's Newsprint

Under the Newsprint Policy announced in 1984, the NP off-take from the indigenous output of four mills was raised to 65 per cent from the earlier 50 per cent. The price of indigenous NP (48 grammage) was higher than the manipulated imported NP supplied by STC. In the case of MPM's NP the price compared to imported NP was as follows:

COST OF NEWSPRINT (*per tonne*)

	MPM	Imported
April 1983	Rs. 7,200	Rs. 5,900
April 1984	Rs. 8,240	Rs. 5,750
May 1984	Rs. 6,900	Rs. 6,120

Another issue, which disturbed publishers during this period, was the lack of uniformity in the levy of Central Sales Tax (CST) by NP-producing indigenous mills. MPM was levying CST at 4 per cent in 1984, but when it was pointed out that the Kerala NP unit which had just gone into production at that time, was not doing so, MPM

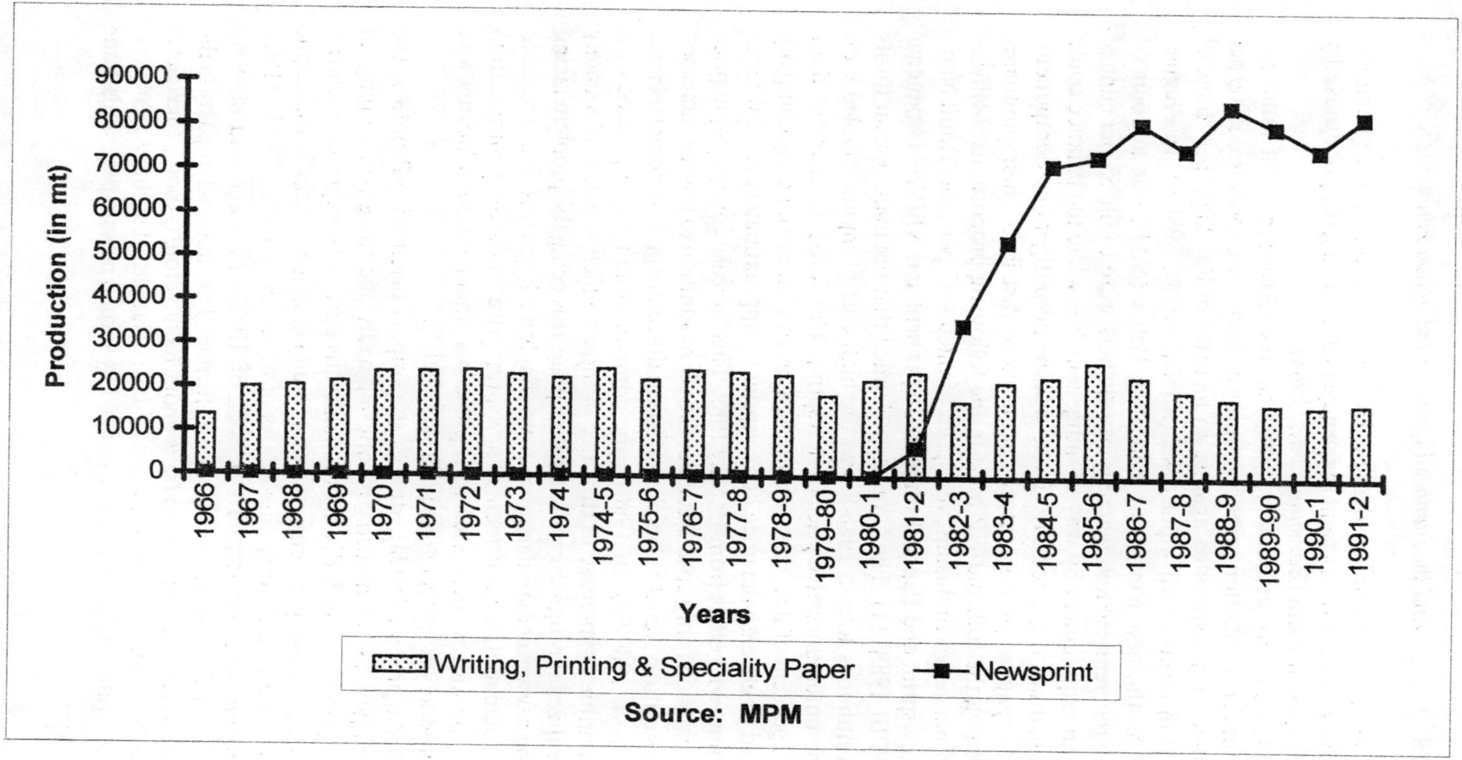

Figure 6.3: Production of Paper and Newsprint (in mt) from 1966 to 1991-2

promptly discontinued the tax. However, it also stopped giving the same quantum of discount to the buyers thus raising the price of NP by an equal amount (*Economic Times*, 1984).

It was only in January 1988, that the Government of India announced a retention price permitting an increase of only Rs. 1,000 per tonne of newsprint as against an increase of Rs. 2,000 per tonne of NP in terms of all input costs (*Annual Report*, 1988-9). In August, 1988, the new Newsprint Allocation Policy (NAP) was announced for the year 1988-9, whereby newspapers were to lift 65 per cent of their requirement by way of indigenous NP, while the balance could constitute imported NP. Even within the indigenous component, newspapers were statutorily required to distribute their purchases among the four old NP mills in the following proportions: MPM—17 per cent, Hindustan Newsprint Limited—17 per cent, Tamil Nadu Newsprint and Papers Limited—16 per cent and NEPA—14 per cent (IDBI 1988: 11). The Company claimed that the main reason for its continuous loss in 1988-9 was the high cost of inputs coupled with the unremunerative price of newsprint. The newsprint accounted for 70 per cent of its total turnover. Against this background the company had to increase the price of newsprint by approximately Rs. 3,000 per tonne with effect from 19 May 1989. This increase covered input price escalations only up to December 1988. In addition to this the increased price also covered only 7.53 per cent of the return on investment (*Annual Report*, 1989-90). In 1990-1, the company bitterly complained of a continuous increase in the price of inputs which was not adequately realized through the remunerative price, this being the problem facing the newsprint manufacturers. Accordingly it felt that this objective could be achieved by adjustment of newsprint price for the full neutralization of periodical increases in input costs under efficient norms and production levels (*Annual Report*, 1990-1).

At present (1994), due to the import-oriented NP policy, the indigenous NP industry is suffering badly and there is 'dumping' of imported NP in India at unremunerative prices by exporter countries at Rs. 400-40 per tonne, the end result being that stocks in the public sector NP mills are piling up. As of today, these mills (TNPL, HNL, NEPA, and MPM), have a stock of 10,000 mt. each and have no place left in the mills to stock any further. The financial impact on the produced NP not sold so far is about Rs. 75 crore (Anon., *Open Letter to Prime Minister*, Government of India, dated 17 February 1994).

'Dumping' of NP is encouraged by Indian newspapers. This same newspaper industry does not lift the indigenous NP quota when it has to, preferring to buy imported NP, the result being unused stock of NP in godowns. In this scenario, where newspapers and dealers who prefer importing NP because it is cheaper are very much aware of the grave danger that our country faces, when the international business cycle takes a turn upwards towards recovery resulting in increased prices of imported NP. The newspapers in such a situation would then be forced to buy imported NP at exorbitant prices if they should throttle the indigenous NP industry (Representative of Indian Newsprint Manufacturers Association, personal communication, February 1994). For details see Ch. 8, HNL.

Raw Material

In 1976, the G.M. of MPM, B. Mohandas, said that the company had already gone in for a mixture of bamboo and several other species of wood in order to conserve the scarce and more expensive bamboo. The daily supply of 250 tonnes of bagasse from the proposed sugar plant would strengthen the raw material position. The bamboo thus conserved would be diverted towards newsprint manufacture (*Business Standard*, 1976).

The paper mills initially had a captive forest cover of 100 km radius with Bhadravati as the centre, which was totally decimated. It then spread its tentacles to neighbouring districts Chikmagalur, Coorg, etc., to harvest bamboo, which were also reduced to naught as mentioned in the earlier section. An intensive campaign was launched by the company to grow fast-growing species required to produce raw material (*Financial Express*, 6 June 1976). It has its own captive plantations.

In 1979, the consumption of bamboo and hardwood by the company was 64,800 tpa which was then expected to increase to 1.25 lakh tpa in addition to the eucalyptus requirement of 90,000 tonnes.

In the same year, to ensure continued supply of raw materials, MPM obtained a lease for 75,000 acres of forest land from the Karnataka government. Bamboo and eucalyptus are cultivated by the company in this captive forest. This is discussed in Section 2.

But the company's forest authorities seem to be unhappy as they claim that 40 per cent of the plantation capital is spent on merely guarding and protecting the plantations (G.S. Kariayappa, Chief

Forest Officer Research–MPM, personal communications, February 1993).

In addition to this, the company has to pay lease rent to the state government at the rate of 12.5 per cent of the yield according to the agreement. We were told that 10,000 tonnes of wood was given as lease rent in 1992.

MPM claimed that during the harvesting of plantations, the lops and tops were taken away by the locals free of cost which they could either use for fuel needs or sell in the market. Apart from this, it was stated that the company distributed 0.5 tonne of firewood on ration cards for every five-member family at the rate of Rs. 200 to 300. It was felt that villagers could afford to buy at this rate as the wood stock lasted them for two to three months (J. Chandra, IFS, personal communication, March 1993).

When MPM embarked on the plantation scheme it met with resistance from the locals who had been using the land for grazing. The company convinced them that the area would be available for grazing after a year when the trees would grow 10-15 ft in height. Similarly 5 per cent of the area was kept for fodder growth (personal communication, March 1993).

Augmenting Raw Material: Sugar Project

The company has put up a sugar plant in the premises of its mill. A letter of intent was received in December 1978 for a sugar plant with a capacity to crush 1,250 tonne of sugar cane per day which was then converted into an industrial licence during 1979-80. The government being approached to enhance the capacity to 2,500 t.p.d., in March 1981, a letter of intent to this effect was issued. Implementation of the project was completed and trial production commenced in June 1984. The total cost of integrated bagasse bulk handling, storage and pulping system was estimated at Rs. 8.20 crore (*SEOD*).

But the *Annual Report* of 1985-6 states that the project implementation was expected to be completed by September 1987. It was further estimated that 3 lakh tonnes of sugar cane crushing in the 1986-7 season would give about 87,000 tonnes of mill wet bagasse which would replace about 43,500 tonnes of bamboo and yield about 15,000 tonnes of cultural paper. Besides, depithing of mill wet bagasse produced about 30 per cent by weight of pith, which could be used in boilers as fuel for generating steam, thereby reducing the overall requirement of coal to the extent of about 11,000 tonnes/annum (*Annual Report*, 1985-6).

Contradictory to this, the *Annual Report* of 1986-7 states that the company generated about 80,000 tonnes of mill wet bagasse and production of paper from bagasse was likely to commence in the later part of 1987-8. The company finally started producing paper from bagasse only in 1987-8 (*Annual Report*, 1987-8).

Production of bagasse improved in 1988-9 and the plant achieved two-thirds of its rated capacity in April 1989. In 1989-90 and 1990-1, 3,000 mt of sugar cane were crushed without adding any balancing equipment. During the same year the company introduced high quality seeds in its seed multiplication programme. The company's own farm operations commenced to enable cane development activities like research, adoptive trials, seed multiplication programmes and demonstration of proven techniques for the farmers. The G.M. of MPM said that, as a part of its conscious policy, the paper mill would open up a remunerative outlet for the sugar cane growers who have been exploited for so long. The company proposed to organize the growers in cooperatives so as to meet the captive cane requirements (*Economic Times*, 25 December 1979).

Effluent Treatment and Pollution Control

An interesting feature of the MPM project is said to be the regulated use of effluent water for cultivation of various crops. This statement, however, is questionable (for details see section on 'From Effluents to Affluence' in Tamil Nadu). It was claimed that for two years, 1977-8, experiments were conducted on growing groundnuts, jowar, maize, and ragi in three different ways—with 100 per cent effluent water; pure water and effluent water mixed in 50:50 proportions, and entirely pure water. Since the water which flowed out of the factory was bluish in colour, it created a panic situation if allowed to flow into the river. Hence it served the purpose of irrigation (*Economic Times*, 1979).

In its *Annual Report* of 1990-1, the company states that it had been active in maintaining a healthy and clean environment. The conversion of boilers along with electrostatic precipitators and high-level chimneys, recycling of lime sludge and extensive plantations were steps in this direction. It further specified that in the matter of effluents MPM always adhered to the prescribed standards through substantial investments. Additional works like creation of fish ponds, recycling waste water within the plant and treatment of solid wastes were being undertaken with free technical assistance from the Government of the Netherlands (*Annual Report*, 1990-1).

The *Annual Report* of 1991-2 states that the technical assistance under Phases I and II programmes of the Dutch government assistance, relating mainly to environment management in pulp and paper, water conservation, solid waste disposal, etc., was completed. The Netherlands government had also offered to continue technical and financial support during Phase III (*Annual Report*, 1991-2).

For installation of the newsprint mill, a loan was taken from an American Bank. Last but not least, the additional quota of raw materials that MPM buys from social forestry programmes are funded by the World Bank.

SECTION 3 (B)

West Coast Paper Mills (WCPM)

It is not clear who had initial control over the West Coast Paper Mills (WCPM). Local loyalties are divided between Somani and to some extent, Bangur. According to a WCPM official, however, the project of West Coast Paper Mills was conceived by A.D. Chaudhari of Karnataka in the 1950s. But he did not have enough resources to set it going. He therefore scouted round, finally at the advice of G.L. Mehta and the newly set up ICICI, he approached the industrial house of G.D. Somani who agreed to take it on. The Somanis set up the mill in 1956, but the major shareholder with 62.5 per cent equity was Bangur who merely provided finance from 1959 till 1973. Somani had a similar type of arrangement with Bangur in other industries such as Shree Niwas Textiles and Digvijay Cement to name a few. The managerial talents were Somanis till 1974, when the Somanis were quietly displaced. G.D. Somani died a heartbroken man (H.K. Somani, private communication, May 1994). The staff of WCPM still speak fondly of the Somanis. The Somanis were forward-looking but Bangur was feudal in outlook (Murthy 1973). The Somanis hail from Rajasthan but their industrial house is Bombay-based.

The Bangurs are Maheshwaris from Didwana in Rajasthan. The information about this family is scanty. The founder is believed to be Mungeeram Bangur who established himself as a banker, property dealer, jute trader and share broker in Calcutta in the 1920s. The

Bangur group came up at the end of the Second World War. They acquired industrial enterprises that formerly belonged to the British, and some to Indians as well.

From the 1920s through the 1940s they had registered a small bank and a large number of finance and investment companies, mainly in the former princely states of Rajasthan. During the 1930s and 1940s, the Bangur family is believed to have made a fortune out of trading in urban land, building up a reputation as skilful dealers in the buying and selling of jute and shares.

Their only industrial ventures were acquisition of Shree Niwas Cotton Mill in Bombay in 1934, setting up another one, Shree Umaid in Rajasthan in 1944, and a cement company Shree Digvijay, in Saurashtra in 1944. Soon after the end of the Second World War, Bangur acquired a number of jute companies from European owners. Later in 1954, the group took over the agency house of Kettlewell Bullen. As a result of investment operations, the group also acquired during the 1950s a substantial interest in several companies under the managing agency of Andrew Yule, as well as in Bengal Paper, a Balmer Laurie company. The interest in Bengal Paper was acquired in collaboration with Dhanuka and culminated in the takeover of the management, a development which was not foreseen by Balmer Laurie (Hazari 1967: 143-5). Did Somani suffer a similar fate? Hazari lists Bangur as the third largest group.

The company was allotted 200 acres of land near Dandeli on a 30-year lease for the construction of its mill by the Bombay government in June 1956, an area which was then a dense impenetrable forest (Kulkarni, personal communication, September 1993). Apart from this it had good water supply from the perennial Kali river and electricity from the Jog grid.

The Rs. 8-crore mill was located amidst the bamboo-rich forest of western ghats in the northern part of Uttar Kannada (UK), now a district in Karnataka. The factory started with an authorized capital of Rs. 5 crore. The equity distribution since 1960 was as follows.

Capital Structure of WCPMs

All shares were issued for cash. Promoters, directors and their friends took up 10,000 preference and 50,000 equity shares. The remaining 50,000 preference and 50,000 equity shares were offered for public subscription through a prospectus in August 1955.

In 1960, 30,000 rights preference shares were issued at par in the proportion 1:2, and 50,000 rights equity shares issued at par in the proportion 1:2.

Between 1961 and 1962 a further 420 shares were allotted out of the 1960 issue.

In January 1966, 50,000 bonus equity shares were issued in the proportion 1:3.

In February 1970, 40,000 bonus equity shares were issued in the proportion 1:5. 60,000 rights equity shares were issued at a premium of Rs. 35 per share in proportion 1:4.

In 1973-4, 60,000 bonus equity shares were issued in the proportion 1:5.

In 1975-6, 40,000 equity shares were issued at a premium of Rs. 35 per share to the financial institutions (IFCI, LIC, and IDBI) on conversion of lease.

In 1979-80, 10,000 equity shares were issued at a premium of Rs. 35 per share to UTI during 1976-7 on conversion of part of the debentures issued to them.

In 1984-5, 90,000 6 per cent preference shares were cancelled and 90,000 12 per cent non-convertible debentures of Rs. 100 each issued in lieu thereof.

In 1991-2, 4,10,000 rights equity shares were issued (preference Rs. 30; 1:1).

DISTRIBUTION OF EQUITY SHAREHOLDINGS OF WCPM
(as on 29 Sept. 1992)

Category	No. of shares held	% to total
I. Foreign Holdings	3,232	0.36
II. Govt./Govt. Sponsored Financial Institutions	2,32,706	26.03
III. Bodies Corporate (not covered) under I & II.	3,73,851	41.81
IV. Directors and their Relatives	11,826	1.32
V. Top 50 Shareholders (not covered under I, II, III, IV)	25,052	2.8
VI. Others	2,47,401	27.68
Total	8,94,068	100.00

Source: Stock Exchange Official Directory.

Additional 61,500 shares were allotted to retain over-subscription, and another 12,500 shares allotted to employees.

The construction of the mill took three years and in 1958 with a capacity of 18,000 tpa it started production of printing, writing, wrapping and kraft papers with bamboo as the sole raw material for pulping.

The Raw Material Question

The manager (forest operation) told the team that even in the 1950s the forest was rich in bamboo.

With the setting up of WCPM, using bamboo as the main raw material, the company approached the Bombay government for concessions in price. In August 1955, the company entered into a 30-year lease agreement with the Government of Bombay for its annual supply of bamboo from 65,000 ha of forests surrounding the mill (NCAER 1970: 51). The agreement gave WCPM the right to extract the material from the forest in accordance with the prescriptions of the working plans.

What is important to note here is the method of calculation of royalty for paper mills in general. For WCPM the concession rate was Rs. 3.75 per tonne of paper. It is interesting to note that the bamboo extracted was not weighed in tonnes. The entire forest material went through different production processes and the final output of paper was taken as a measure to pay the royalty. The sulphate process that the forest-based paper mills in India use utilizes only 35 to 40 per cent of the cellulosic raw material, the rest being wasted. The tremendous loss to the Forest Department is due to this faulty system of calculating royalty. It works out to be Re. 1 per tonne.

Within a year of production, bamboo flowered gregariously in waves and died. Since paper machines were fabricated only for bamboo, WCPM took a keen interest in regeneration of bamboo (details in Section 2).

The original lease agreement with the state government, which was revised in the year 1973, had assured sustained supply of raw materials to the mills. In January 1989, the state government cancelled all agreements of raw material support to the mills because of fast-depleting forest resources.

From the time bamboo flowered in 1962, the company had been aware of the grave danger it faced, and accordingly diverted its attention towards eucalyptus wood to make up for the shortfall of bamboo. The use of hardwood (eucalyptus) started with 10 per cent as a total percentage of raw materials used. This steadily went up with the corresponding decline in availability of bamboo.

The mill complained that the government did not allow it to grow plantations on degraded land. The only alternative was then to approach farmers to grow pulpable species on their private lands. In the process, the mill authorities discovered *inami* lands, the lands given to villagers by the British as *inams* (prizes). These lands were uneven and not very fertile. In 1962 the company leased in 5,728 acres of *inami* lands for 30 years and planted eucalyptus on them. Since the land was infertile and of bad quality, the rent decided upon by the mill was a meagre Rs. 9.50 per acre per year. The mill authorities claim that this was a gamble. These lands were 56 km away from the mill at Kulahalli in Dharwad district. The plantations named 'Banjivan' in honour of G.D. Somani, came up very well and the per annum yield was 1,71,000 tonnes. A report on the visit to these plantations has been included in Section 2.

The plantations are due for renovation but the company complains that the government is preventing extension of the lease as the land now belongs to the state government under the Karnataka Land Reforms Act, Inam Abolition Act, etc. This issue was brought up in 1978.

When the state government brought its claim over the land, the mill authorities stated that they approached the government to grant permission to continue with their plantations on these lands. In turn the company would pay Rs. 9.50 per acre per year as rent to the government. The landowners filed a writ petition in the High Court of Karnataka and obtained an interim stay in 1978. The company also gained a lot over this argument about the ownership of land. Now it neither pays the rent to the state government nor to the inamdars, but continues to draw large amounts of eucalyptus wood from these lands.

In addition to this, the eucalyptus grown in the entire Karnataka state and by the Karnataka Forest Development Corporation before 1982 was allotted to Birla's Harihar Polyfibres, which produces viscose staple fibre (see Figure 6.1). WCPM agitated against this and asked for equitable distribution of eucalyptus wood from the state. The Government of Karnataka informed WCPM that it was because of the contract with Harihar Polyfibres that all the wood was supplied to it. WCPM filed a petition in the High Court. This conflict went on for two years, ending in a mutual decision that 47.5 per cent of eucalyptus would go to WCPM and 52.5 per cent to Harihar.

The present consumption of hardwoods by WCPM is 94 per cent while the remaining 6 per cent is dead and decayed bamboo which it gets from the FD. The mill authorities complain that as the extraction is incompetently and inefficiently done by the FD, the yield of dead bamboo is very low—only 12,000 to 15,000 tonnes. This bamboo is brought into the auction depot, West Coast bidding the highest price, as its competitor MPM cannot do so due to un-economic transportation charges (A.S. Bhat, September 1993, personal communication).

The mill also claims that it purchases eucalyptus wood from private farmers from Bangalore. These lands are usually open lands and belong to absentee landlords. The landlords fear that their lands will be encroached upon, therefore to keep the land occupied they raise plantations which do not require continuous attention, but produce enough to fetch a good price.

The company is planning to set up a bagasse pulping unit and a sugar mill to meet its future raw material needs. Company officials claimed that sugar cane was available in plenty, but due to low demand in the area the farmers have to travel a great distance to sell their produce. Presently the WCPM cannot afford to initiate action in this direction because of lack of resources. It has yet to recover from its recent 1993 renovation programme. In addition to this, the company plants to invest its resources for bamboo plantations on a joint basis with the Karnataka Forest Development Corporation in the Belgaum district—a high rainfall area. ([Superintendent] and a lower level official of FDC of Karnataka, private communication, October, 1993). Figures 6.4 and 6.5 show the quantum of material supplied to WCPM by the Forest Department and Forest Development Corporation, and the royalty rate charged.

Mill Operations

The commercial production of WCPM started in 1958 with two paper machines, one MF and the other an MG imported from Germany, with an installed capacity of 18,000 tpa. The products manufactured were writing, printing and packaging papers. The third machine, an Italian-made MF was added in 1972 and production began to rise and the product line changed. Production in 1965-6 was 34,123 tonnes.

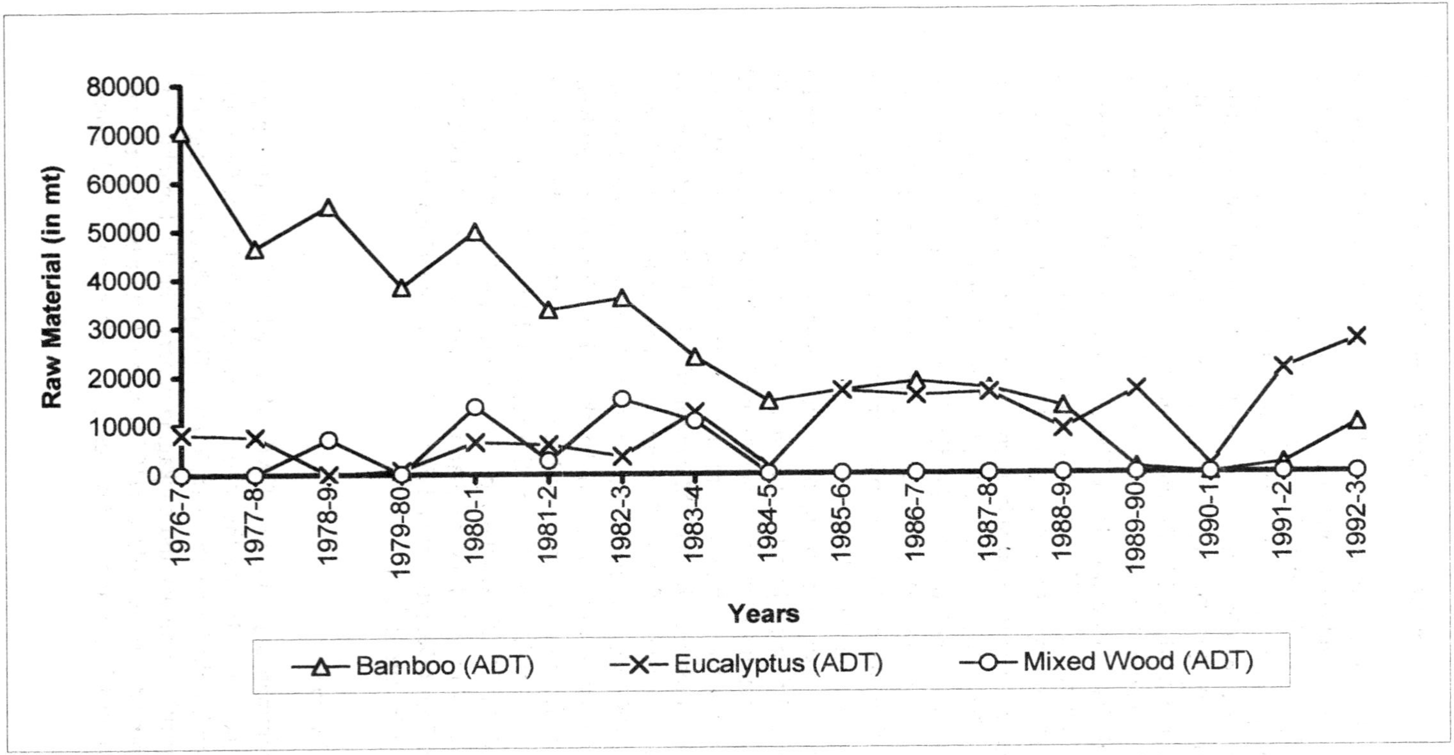

Figure 6.4: Raw Material received by WCPM from Forest Department and Forest Development Corporation 1976-7 to 1992-3

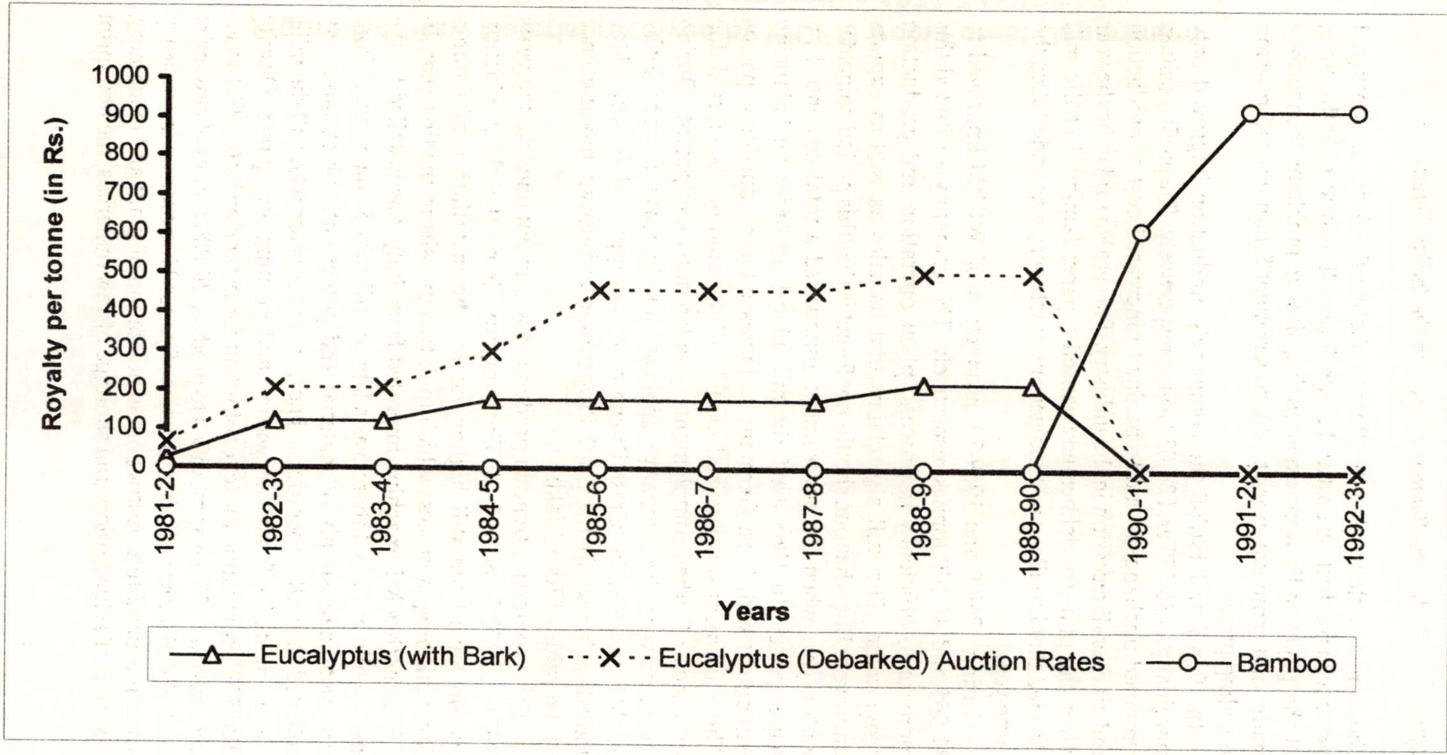

Figure 6.5: Royalty per tonne for Raw Material charged to WCPM from 1980-1 to 1992-3

The company claims that it suffered a slight setback due to continued increase in production cost and controls on selling prices. At this time, all the paper mills were experiencing similar problems, hence they made joint representations to the government for an upward revision in selling prices.

In 1966-7 and 1967-8 the company's production improved only marginally to 35,575 and 36,032 tonne respectively. The profit position remained unchanged, said to be due to increased cost of production and an overall recession in the industry. This is also said to be the year when the industry deliberately caused a paper famine. But the mill claimed that installation of indigenous machinery was undertaken the same year.

During 1968-9 and 1969-70, production of WCPM improved to 37,565 and 41,100 tonnes respectively. The company attributed this increase to efficiency in process, plant operation and production programme. The mill also claims that its sales as well as exports improved. But what one should not forget is that this period coincides with a year when the government announced a total de-control on paper prices. In 1968 itself the mill places an order for balancing equipment with West Germany to increase the production capacity. The equipment was received in 1970-1 but could not be erected due to a lack of other complementary indigenous equipment. During the same year, the company's application to increase the capacity by 15,000 tpa under the 'Crash Programme' was approved. It received the industrial licence for the Crash Programme and also an import licence of the value of Rs. 91 lakh.

This progressive trend in production was maintained for the years 1971-2, 1972-3 and 1973-4. The company claims that the production would have been still higher but for the installation of balancing equipment a process which took two months in the first year. In 1972 another MF machine imported from Italy was installed by the mill, while in 1973-4 there was a labour strike for 50 days, apart from an inadequate supply of material and power cuts which adversely affected the mills' operations. These were the years when all the mills again created a famine in writing and printing levy paper. Despite WCPM's complaint, the letter of intent for expansion by 60,000 tonnes received in 1972-3 was modified to produce 30,000 tonnes of newsprint and 30,000 tonnes of other varieties of paper. However, owing to heavy capital expenditure, restriction on utilization of plant and machinery, fixation of selling prices and shortage of raw materials, the newsprint project was dropped the very next year, 1973-4. All the mills had shifted their position with regard to production and sales, showing considerable improvement, which absorbed the all-round rise in cost of production.

Though GO Control of Production and Price (1974) was brought in, production fell the next year. The company pinpoints the reason to change in grammage substance in compliance with the Paper (Control of Production) Order effective from 1 August 1974. The political scenario just before the emergency was grim. Every unit complied, creating a glut in writing and printing papers. WCPM paints a grim picture for the years 1975-6 and 1976-7, claiming poor offtake of paper due to sluggish market conditions, reduction of selling prices from time to time and power cuts. In 1976, with a glut, paper industries bargained with the government and raised their prices. The same year, the new paper machine showed signs of stabilization and the company was happy that the low gram weight paper manufactured was well received in the market. The reports however, do not indicate which type of paper was well received (*SEOD*, 1983)

While production rose to 53,338 tonnes in 1977-8 from 51,578 tonnes in 1976-7, sales were lower than in the previous year. The PPI was hiking prices, so that WCPM's value stood higher at 19.68 crore, because of improvement in the sales in the market (*SEOD*).

The paper industry as a whole was on the warpath with the government on various issues such as royalty rates, insistence on levy paper, price control, etc. Production was therefore almost at a standstill. WCPM's production and sales in terms of quantity were lower at 51,618 and 51,775 tonnes respectively. The fall in production was attributed to the non-availability of adequate supplies of raw material, coal and power, and to some extent labour unrest for 1978-9. In the same year a coal-fired boiler was commissioned which helped to increase the mills' own generation of power. During 1979-80, production and sales recorded a steep fall, from 51,618 to 36,739 tonnes due to erratic and inadequate supplies of power and coal which resulted in frequent shutdowns of the paper machines, adversely affecting the production for the year 1980-1, and eroding the profitability of the mill (*SEOD*). Also, there was a strike for 109 days. It is interesting that all the forest-based mills had similar complaints during this period.

WCPM's performance was satisfactory for 1981-2, 1982-3 and 1983-4. High capacity utilization of 108 per cent was achieved leading to high production. The sales also registered an increase from October

1983, following market improvement in the demand for paper. To solve the power problem the mill installed one 1200 kva diesel generating set in 1983-4 (*SEOD*).

In 1984-5 production increased to 75,082 from 64,812 tonnes in 1983-4, despite inadequate supplies of raw material. In 1985-6 working results were unsatisfactory mainly on account of power cuts and mounting production costs (*SEOD*).

The mill produced 11,103 tonnes of white printing paper in 1986-7 which it sold to the government up to 22 January 1987, at a concessional rate which, the company claimed, was the cause of average realization. The Paper Control Order, however, was repealed with effect from 22 January 1987. In the years 1988-9, and 1989-90, production remained at the previous year's level while in the later years better market conditions naturally resulted in improved performance.

Presently WCPM's capacity stands at 60,000 tpa (licensed and installed) producing a larger variety of paper (see Table 6.5).

The corresponding increase in demand for raw material went up to 3 lakh tonnes per annum (production and sales figures are represented in Figure 6.6). In order to augment the working capital resources, the company sold some of its old equipment to various leasing companies for Rs. 200.92 lakh under 'Sale and Lease Back Scheme' (*SEOD*).

The present production of paper is 200 tonnes per day. Out of this, 3 per cent low quality or defective paper is recycled. The pulping

TABLE 6.5: TYPES OF PAPER PRODUCED BY WCPM

1.	Azure laid/wove, Deluxe, Yub-sized
2.	Bible Paper
3.	Dyeline Bass Paper
4.	Copier Paper
5.	Coating Base Papers
6.	Cover Papers, MF, MG, SS, TDL
7.	Cheque Papers, MICR & Non-MICR
8.	Cream Wove
9.	Kraft Papers, MG Plain and Ribbed
10.	Mottling Papers
11.	Parchment Papers
12.	Posters MG, White and Coloured
13.	Printing Papers, Maplitho
14.	Pulp Boards, MF & MG, White and Coloured
15.	Special Varieties

Source: West Coast Paper Mills (13-16 September 1993).

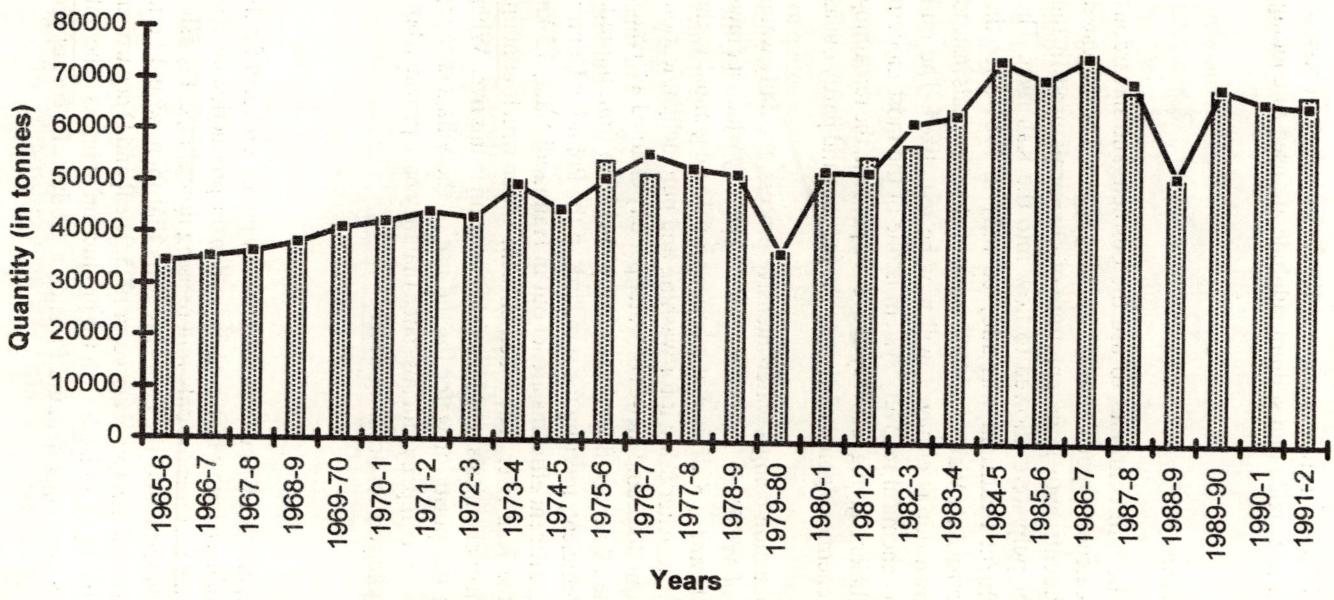

Source: WCPM, September 1993

Figure 6.6: WCPM Production and Sales from 1965-6 to 1991-2

process used for manufacture of paper is Kraft (Sulphate) process, in which sodium hydroxide and sodium sulphide are used as cooking chemicals.

WCPM's Effluents

Even today the mill continues to use the obsolete and hazardous chlorine and calcium hypochlorite as bleaching agents. The untreated corrosive effluents with large amounts of chemicals used in the production of pulp were allowed to flow into the Kali river. The statutory effluent treatment was adopted only in 1981 after the Pollution Control Act (Water Act) 1981 came in; the chemicals instead of being thrown out as effluents, will now be re-utilized. The mill claims that its chemical recovery system is one of the most modern and highly efficient, giving recovery of over 91 per cent, the remaining 9 per cent of chemicals going partly into the product and partly out as effluents.

Paper production being a water-intensive process, the daily water requirement of WCPM is 75,000 cubic metres per day which it draws from the upstream of Kali river at a concessional rate granted in the lease agreement. After using it for washing and pulp delignification, the mill lets out 300 cubic metres of water per day claimed as treated effluents. The 30 acres effluent treatment plant, 2 km from the mill, is said to be efficient, meeting the standards of the Pollution Control Board. The treatment effluents are let out in Halmaddi Nala, 0.5 km away from the Kali river. The black effluent water which contains some amount of chemicals gets diluted downstream in this river. When asked about the purity of water for the use of the villagers living downstream, the company officials stated that they have provided bore wells to the villagers for drinking water.

Lime Sludge

Lime sludge, another waste from the recovery plant, is reburnt for the production of fresh lime which is again used in the production process. WCPM claims that it is the only mill re-using lime sludge. Fly ash is dumped as landfill, and the company fears that there would be no ravines to fill in another two years, i.e. in 1995. The R and D department is therefore trying to utilize fly ash for production of bricks. The only problem is in creating a demand for such a product. Its cost of production and transportation would be very high in comparison with local bricks.

Coal

Coal, another major raw material, is procured from collieries in AP, MP and Maharashtra. The company complains that it is allotted lower quality coal while the first priority for best coal is given to the steel industry.

Power

The mill has its own captive thermal power generation system which generates 50 per cent power, the other 50 per cent being supplied by the Karnataka Electricity Board. The total energy requirement is about 1 crore units per month.

Product Mix

WCPM diversified from the traditional writing and printing paper into several varieties in the late 1960s. The varieties of paper being produced continue to be marketed as Sudarshan Brand. The total production of paper includes writing, printing, bond, manila, postal, chrome and art paper. With completion of its modernization programme in May 1993, the mill claims that it can produce quality papers like light-weight paper including surface-sized papers for magazine grade and other printing requirements. The major consumer of WCPM's product is WD and HO Wills Cigarettes.

Summary and Conclusion

Mysore was once a native state. Under the stewardship of wise administrators the state undertook a series of progressive steps towards industrialization. The first paper mill was set up in 1937, inevitably coinciding in time with those set up by Birla, Dalmia, Bajoria and Singhania. It was only when the political situation stabilized that Bombay-based Somani entered Dandeli. The two mills and Birla's Harihar Polyfibre between them swept away the bamboo forests of Shimoga and Uttar Kannada, as well as the forests of neighbouring districts. Somani wisely listened to the advice of his forest manager,

A.S. Bhat, who undertook to revive bamboos and even protect them. West Coast Paper Mills is admired for this act. Somani was quietly eased out by his partner Bangur, who neither cared for the bamboo forests nor for modernizing the mill, till the second generation took over.

Meanwhile, the Government of India, faced with the acute problem of shortage of paper, particularly newsprint, allowed the state-controlled MPM to modernize and produce newsprint. The State Forest Department helped MPM to set up vast plantations of eucalyptus and *Acacia auriculiformis* to feed the industry. ODA of UK collaborated, perhaps even initiated the process. The main question raised and discussed here is: Can plantations replace natural forests? Again, if the species do not match the soil, will this lead to further degradation? Can such experiments on a large-scale without pilot studies be worthwhile? Does the natural revival ofbamboo in all its glory establish our hypothesis that after a ban on felling was imposed over-felling leads to destruction of the bamboo forests and even defensively induce gregarious flowering?

Birla, known for his total unconcern for forests, consumed the vast areas gratuitously given to him in Kerala, and moved on to neighbouring Karnataka to lay waste the forests there. He even 'induced' the state forest department in KFDC to cultivate eucalyptus as a joint-sector venture, much to the annoyance of WCPM, the rival for the raw material. Birla's Harihar Polyfibres produces pulp for rayon, not for paper. It was finally the local farmers who through public litigation put a stop to Birla taking advantage of the State Forest Department and the Forest Development Corporation and growing eucalyptus for HPF. The political economy of paper and the political economy of rayon are somewhat different, but only the former is discussed here. Even now, the countryside, the roadside in cities and towns of Karnataka are lined with eucalyptus and acacia, sometimes five deep.

Recommendations

1. It is imperative that all plantation forestry especially with exotic pulpwood trees must cease.
2. Attempts must be made by the Forest Department to find out whether clear-felled areas of pulpwood trees in heavy rainfall areas would be able to revert back to natural forests. Succession has been noted in WPM's Kulahalli plantations and in Kerala. Bamboo areas of Karnataka also pointed to this trend.

3. Appropriate species of bamboo should be planted in bamboo-denuded areas. The felling cycle of bamboos should be lengthened and bamboos harvested with care for MPM and WCPM.
4. Farm forestry with exotic trees should be replaced by known cellulose-yielding hibiscus varieties on farms. Farmers must be helped to grow these varieties, perhaps as rotation crops. They need the help of agricultural scientists and financial institutions.
5. The state-owned MPM must be helped to grow and bureaucratic interference should cease.
6. Indian industry must be protected from foreign competition. The benefits of such protection should be shared by the consumer; the state should take on the responsibility to have this implemented.

NOTES

1. KFPC was still raising cocoa (1,056 ha), rubber (4,443 ha) and *bunsera* plantations. The expenditure incurred by KFPC till 1984-5 was Rs. 522.4 lakh.
2. The mill operates its plant for seven days a week on a three-shifts-a-day basis. By and large the utilization of pulp is determined by the operation of paper machines. In genereal the capacities in the pulp stage are in excess. Loss of production due to downtime in the pulp mills, if any, gets reflected in the downtime of the paper machines. Consequently, the on-stream efficiency of paper machines is indicative of the overall efficiency of the mill.

REFERENCES

Abhishankar, K. 1975, *Karnataka District Gazetteer, Shimoga*, Government Press, Bangalore.

——— 1985, *Karnataka District Gazetteer, Uttar Kannada*, Government Press, Bangalore.

Ananthnarayan, N. 1979, 'Mysore Paper Launches Expansion Project', *Financial Express*, 7 May.

Anon. 1938, Mysore and Modern Industries, Section I, Industrial Development in Mysore State.

——— 1976, 'Mysore Paper Mills to Expand', *Business Standard*, 5 November.

——— 1978, 'Mysore Paper Mills Expansion Apace', *Financial Express*, 6 June.

——— 1979, 'Mysore Paper to Roll Out 75,000 Tonnes Newsprint by June', *Economic Times*, 25 December.

——— 1979, 'Bhadravati Newsprint Project—A Giant Step Towards Self-reliance', *Economic Times.*

——— 1982, 'IDBI not Keen to Fund MPM Project', *Business Standard*, 6 March.

——— 1983, *Bombay District Gazetteer*, North Kanara, Bombay.

——— 1984, *Bombay District Gazetteer*, Dharwar, Bombay.

——— 1984, 'Newsprint Muddle, Publishers Costs will Rise by Rs. 12 Crore', *Economic Times*, 3 August.

——— 1993, *Journal of Industry and Commerce*, vol. XXX, no. 5, May 1993.

——— 1994, 'Open Letter to the Honourable Prime Minister', 17 February.

Bergbie, L.F.B. 1910, *District Gazetteers of Central Provinces, Chanda*, vol. I A, Pioneer Press, Allahabad.

Bhat, A.S. 1993, *A Note on Raw Material*, WCPM, Dandeli, 13 September.

Chandran, M.D. Subash 1984, 'Ecosystem Studies and Management in Coastal Uttar Kannada', *Technical Report*-3, Indian Institute of Science, Bangalore.

Chandrashekar, D.M. and B.V. Krishnamurthi 1987, *Economic and Political Weekly*, 13 June, no. 6, Bombay.

Chaturvedi, A.N. 1976, 'Eucalyptus in India', *Indian Forester*, vol. 102, no. 1.

——— 1984, *My Forests*, vol. 20, no. 4.

FAO 1949, *Fourth Meeting of the Technical Committee on Wood Chemistry*, held at Brussels.

——— 1954, *World Pulp and Paper Resources and Prospects*, New York.

——— 1981, *Forestry and Rural Development, Forestry Paper*, no. 26, Rome.

——— 1980, *Pulping and Paper Making Properties of Fast Growing Plantation Species*, vols. I and II, *Forestry Paper*, 19.1 and 19.2.

——— 1985, *The Ecological Effects of Eucalyptus, Forestry Paper*, 59.

——— 1992, *Mixed and Pure Plantation in Tropics and Subtropics, Forestry Paper* 103, Rome.

——— 1993, *Assessment of Forestry Project Impact, Forestry Paper*, 114, Rome.

Francis, W. 1872 and 1907, *Madras District Gazetteers, Visakhapatnam*, Government Press, Madras.

Gadgil, M. and N. Prasad 1984, 'Indian Bamboo Species: The Vanishing Bamboo Stocks', *Biotropic* 16 (3).

——— 'Bamboo in the Rural Life of the Hill Tracts of Western India', *Proceeding of Indian National Science Academy*, B 51, no. 51.

——— 1978, 'Vanishing Bamboo Stocks', in *Commerce*, 17 June.

Gadgil, M. 1989, 'Forest Management, Deforestation and People's Impoverishment', *Social Action*, vol. XXXIX, no. 4, Oct.-Dec.

——— 1991, 'Science and Natural Resource Management', *Seminar*—The Monthly Symposium, New Delhi, India, 389.

Gibson, A. 1866, *A Handbook of Forests*, 1857-60, Bombay Government Press.

Guha, R. and M. Gadgil 1989, 'State Forestry and Social Conflict in Industry', *Past and Present*, 123.

Hazari, R.K. 1967, *Structure of the Private Corporate Sector*, Asia Publishing House, Bombay.

Heweston 1962, 'Forests of Chanda', in H Champion and F. Osmaston, *Stebbings Forests of India*, vol. IV, Oxford University Press, London.

Huberman, M.A. and C. Purukayastha 1951, *Australian Forestry*, vol. XV (FRI Documentation Centre).

IDBI, 1988, 'Paper and Paper Products—Newsprint Allocation Policy', *Industry News Digest.*

Kadambi, K. 1949, 'Ecology and Silviculture of Dendrocalamus strictus', *Indian Forester*, vol. 75.

Kallarackhal, Jose and C.K. Soman 1993, 'Water Use of Selected Indigenous and Exotic Species', Plant Physiology Division, KFRI (unpublished).

Karnataka Forest Department, *Annual Reports*, 1985-6 to 1992-3, Government of Karnataka, Bangalore.

——— 1987, Statistical Brochure, Government of Karnataka, Bangalore.

——— (n.d.), *Forestry in Karnataka*, Bangalore.

Karnataka Forest Development Corporation Ltd., *Annual Reports* for 1971-2, 1984-5, 1986-7, 1988-9 and 1990-1.

Kohli, R.K. 1990, *Allelopathic Properties of Eucalyptus*, Botany Department, Punjab University, Chandigarh.

Kormand, Edward 1991, *Concepts in Ecology*, Prentice Hall, New Delhi.

Kuruvilla, P. 1979, 'No Paper Tiger This', MPM Means Business, *Business Standard*, 7 May.

Lakshmanna 1990, *Bamboo: A Potential Employment Generator; Its Present Status and Future Prospects*, in Proceedings of National Seminar on Bamboo, Dec. 1990, Aranya Bhavan, Karnataka.

Muqhabe, J. 1989, *Forest Ecology and Managament.*

Murthi, R.C. 1979, 'Bangur Somani Split Complete', *Business Standard*, 1 January.

Muthanna, M.A. 1955, 'Mysore Forest Department, A Short History', *Indian Forester*, vol. LXXXI, no. 7, Dehradun.

Mysore Paper Mills, *Annual Reports of 1985-6 to 1991-2*, Bhadravati.

——— 1993, MPM *Captive Plantations: A Note*, Bhadravati.

Nair, V.S.K. 1988, 'Forestry Programme of MPM Limited', *Proceedings of Seminar on Raw Material for Paper Industry*, Indian Paper Mill Association.

Prasad, N. 1981, 'Ecology and Management of Bamboo Resources of Karnataka', Unpublished Ph.D. Thesis, Centre for Ecological Studies, Indian Institute of Science.

Reddy, Yellappa and T.K. Karwari, 'Eco-Restoration and Dryland Farming', *Karnataka State of Environment*, vol. V, Bangalore.

Sagreiya, K.P. 1977, *Indian Forester*, vol. II, Dehradun.

Sankaran, K.V., M. Balasundaram, T.P. Thomas, M.P. Sujatha 1992, 'Litter Dynamics, Mycrobial Association and Soil Studies in Acacia auriculiformis Plantation in Kerala', unpublished, KFRI Project 126/89.

——— 1993, '*Albezzia Parserian*, *Eucalyptus tereticornis* and *Tectona grandis* in Kerala', in *Forest Ecology and Management*, 56, Amsterdam.

Shiva, Vandana, Bandopadhyaya 1981, *Economic and Political Weekly*, Bombay.

Shyam Sundar 1987, 'Social Forestry in Karnataka', *My Forests*, 1987, vol. XXIII, no. 4, 9.

Shyam Sundar and Parmeshwarappa, 1989, 'Social Forestry and Eucalyptus', *Economic and Political Weekly*, 7 January, Bombay.

Stock Exchange Official Directory (*SEOD*), *Papers on Mysore Paper Mills*, Bombay, 1964-5 onwards.

——— *Paper on West Coast Paper Mills*, Bombay, 1964-5.

Thirawat, Sukhum 1956, *Indian Forester*, vol. 82, no. 12.

Von Monroy, J.A. 1961, *Report to Government of India on Integration of Forests and Forest Industries*, India F5, FAO, Rome.

West Coast Paper Mills, 1993, *A Profile*, WCPM, Dandeli.

CHAPTER 7

Kerala

Bamboo for Whom?

Compassionate Keeper
Do tend your bamboo and cherish your Babe
but don't push your children into limbo.
This is the emerald green land
where now,
the sapphire darkens the sky,
and Ka the python once committed suicide
while the grizzly Bear had to hibernate.
Do you get me—oh wise keeper?

M.S.

SECTION 1

MALAYALEE REGION

The state of Travancore-Cochin along with Malabar and the Malayalee majority talukas of south Kanara from the Madras Presidency were together constituted as Kerala in 1956.

Kerala is a small state densely populated, densely wooded, with a potentially good stock of bamboo. It is also the most progressive state in India. This study touches on only three relevant progressive aspects of Kerala. It is the only state in India which has a fully literate population. Logically therefore, printing and writing paper and newsprint are an integral part of such a society. Secondly, since the state encouraged this, it now has a highly conscious labour force. Finally it brought in land reform after the state reorganization. But immediately after the Second World War it condoned encroachments upon its precious forests. T.P. Vishwanathan, retired Chief Conservator took pains to put the situation into perspective for the researcher (1993); land hunger in a small new state with limited arable land and no industries was tremendous, but the state and the forest department took care to see that no incoming family could occupy more than one acre of forest land.

The newly-formed state was under dual pressure, international as well as national. At the formal but invisible international pressure level was FAO, and at the national the formidable pulp and paper industry with Birla heading one of these industries. The centre's animus to both the left and left front governments intensified the pressure. These are not disparate factors, but interlinked in effect.

In Section 2, this study raises a very controversial issue.

A section of foresters and environmentalists firmly holds the view that bamboos and reeds *Ochlandra* species of bamboo should be totally reserved for basket and mat weavers and all the PPI units should be shut down. We hold that such a viewpoint is ahistorical and that a distinction should be made between socially useful and wasteful use of scarce bamboo resources. Felling of timber is outside the scope of the study but touched upon for the first time to get a fuller picture of the changing face of the Indian forests. The first to ravage these forests were the colonizers.

Colonial Legacies

In this section an attempt is made to re-create a picture of the forests of the state of Kerala during the colonial period. This is afforded through the various accounts by colonial foresters.

Malabar district, formerly a part of Madras Presidency, was won by the British from Tipu Sultan in 1792. Though it was formally taken over, no steps were taken to exercise the right of possession whereby large landowners took possession of the forests. In fact, since the dawn of the eighteenth century, the forests of Cochin were under the possession of the feudal chiefs called 'naaduvazhies'. Early Portuguese and Dutch traders extracted large quantities of teak from the forests. A timber syndicate headed by Machonochie was formed in Malabar in 1799. The syndicate progressed for sometime and then collapsed. Till 1799, 10,000 teak trees were brought down the Beypur river (Malabar). It was estimated that 2,000-3,000 trees may have been annually procured [*Millburn Oriental Commerce*, vol. I (1813): 328].

The abundant teak in Malabar attracted the notice of the Bombay government to which the province of Malabar had belonged for some years. The province was transferred to Madras in 1800. A Bengal–

Bombay Joint Commission was set up to put regulations on the felling of teak below 21 inch girth in 1805. At this time there was a deficiency of oak in England. The court of Directors of the East India Company enquired to what extent the King's navy required teak from Malabar. This is the period when interest was aroused in the forests of India. The safety of the empire depended on its 'wooden walls'. A Forest Committee was set up which sent in its report saying that the more accessible teak had been cut and roads would have to be built to gain access to the available teak. They had also over-estimated the capacity of teak to mature. The immediate response to this report was a proclamation, declaring that the royalty right in teak was vested in the Company and a forest officer was forthwith appointed to superintend the forest work, under pressure from the Home government. Captain Watson of the police was appointed as the first conservator of forests on 10 November 1806. In Palghat, the conservator succeeded within a short space of time in establishing a monopoly of all the timber throughout Malabar and Travancore. He felled timber not only from private forests but also from government forests paying revenue to the government. Private timber trade was annihilated.

Reliable accounts on the working of the forests of Travancore area are recorded by Ward and Croner (1817-20). They reported that it was customary to rent each river for its timber (only teak) and this ruined the forests. It was becoming difficult to get teak anywhere except in the far interior. Teak was the only tree monopolized by the state at that time and the other species could be freely felled by the people, except that all wood was subject to a light river duty when transported by river.

By 1823 the growing discontent of the forest proprietors and the timber merchants under the restrictions of the timber monopoly and the outcry of the peasants indignant at the fuel-cutting restrictions came to a head. Under these circumstances, the Governor Sir Thomas Munro recommended the abolition of the conservatorship, which was done the same year.

The landowners took possession and once again reigned supreme. Even government forests were subjected to unrestricted felling by the timber syndicates. This agency, under the supervision of an officer of the navy, made higher payments to local contractors. There was a ruthless exploitation of teak and as a result the British navy faced a shortage of timber supply at the Bombay Naval Yard in 1830. In 1828 Sheffield, the principal Collector of Malabar, brought this wanton destruction of the forests to the government's notice. Also in 1830, the Nilambur Raja brought this ruthless destruction to the notice of the British. The Government of Bombay asked the Navy Board to submit a report on the Malabar forests. In reply, the navy suggested reinstatement of the conservatorship which was duly done in 1831. Again, in a letter to the government Clementon, the principal Collector of Malabar in 1834, condemned all direct interference by government officers. In 1838 the Madras revenue board suggested that there was no need for an independent authority and there should only be a Revenue Officer in charge of the forests.

Stebbings (1920) reacts to this, saying that while government supported the rights of private ownership, it forgot the equal rights of the community which were gravely imperiled because of the destruction of the material required by the far more numerous body of agriculturists, consequently large areas would go out of cultivation owing to the diminishing water supplies. Mukundan and Meher Singh (1991), point out that while the British systematically exploited the timber required for the empire, giving the people rights over the less valuable timber, they also alienated them thereby starting off the process of ecological imbalance.

Forests Cleared for Plantations

It was probably the over-exploitation of the teak and the fear that there would be a shortage soon, that prompted the British to get into plantation activity. In 1840 H.O. Conolly, the Collector of Malabar, put up a proposal for acquiring 670 sq. km of teak to be managed properly. Though he was not allotted the area he had asked for, he was allotted areas elsewhere. Chathu Menon, appointed sub-conservator in 1844, who with Conolly, raised successful teak plantations. Meanwhile, the Forest Department was engaged in working teak areas all over the state. A total of 1,500 logs were transported between 1835 and 1842. In 1844 blackwood and *anjili* were also declared as monopolies along with teak. In 1865, the duty on timber was adjusted, and the sale of wastelands in the hills and low country were allowed for the cultivators to take up lands for plantations of coffee and other products.

Early accounts recorded by Bourdillon (1892), tell of coffee being grown in the lowlands on land given away by the Travancore government and later on sold at Re. 1 per acre. These lands were both in the hills and the plains. In 1873, the government raised the prices to Rs. 10 per acre. There was a rush for these lands. With this, about

39,400 acres of land, mostly virgin forests, were sold to Europeans and a few natives. After 1877 the price of coffee began to fall and the estates began to be abandoned. The terrain was steep, the soil easily washed away and cultivation was not properly managed as a result of which, the 'leaf disease' spread and coffee failed. The European planters took up *Cinchona succirubra* plantations. Tea was introduced by General Cullen in 1876 and its cultivation was extended into 3,000 acres.

Bourdillon (1892) was of the opinion that half of the state was forested and that these forests must have earlier been more extensive, probably covering the entire state.

Commercial Forestry

Meanwhile, forestry activities were being carried out quite intensively. In 1866, felling of trees along the river banks was prohibited by virtue of a proclamation, and again another order forbade hill people and cultivators from cutting any forest trees. There was also a prohibitory order on shooting elephants and restrictions on hill cultivation, but this order was not successfully implemented.

From 1871 onwards, timber was delivered at depots and sold in auctions, and the system of seniorage was abolished. But 1888 saw the abolishing of the depot system and reintroduction of seniorage. In the same year the first Forest Act was passed, declaring Konni as the first forest reserve.

Again the year 1888 saw Punalur Paper Mills being set up. It was built by the Travancore Paper Mills Company on the banks of the Kallada river. The principal raw material was Etta/reed (*Ochlandra travancoria* and *O. rheedii*) collected from the government forests on a long-term agreement on payment of the stipulated seniorage. The main collecting areas were situated 30 to 120 miles from the mill.

In 1896-7 the forest department was reorganized and the forests were divided into four main administrative divisions: (a) Thenmala, (b) Punalur, (c) Ranni, and (d) Konni.

Basha (1990), says that the destruction of forests in Kerala began with the onset of developmental activities in the nineteenth century. Railway tracks from Shencottai to Quilon and Coimbatore to Cochin came up at the expense of the rich, moist, deciduous forests. The fuel demand for the railways was met by the forests of Walayar and Dhoni in Palghat, the teak and other timber requirements for coaches were met from the natural teak forests.

The two World Wars placed heavy demand on the timber in the forests. The most affected were the forests of Parambikulam and Chalakuddy, as special tramways were constructed to transport the timber. The forests on the lower slopes of Malabar were also put to the axe. The Kallai timber market (the biggest in Asia at that time) dealt with timber bought from the private forests of Malabar. The major rivers like the Chaliyar provided natural transport for timber. The Calicut and Cochin ports served as outlets for the export of hardwood. Alongside this activity, deciduous forests of the lower slopes were cleared for plantation crops like rubber, coconut, etc.

Wartime timber needs were met very systematically. A working plan circle was formed in 1926, which worked the areas in Palghat, Nilambur and Wynaad, till then a part of the Madras Presidency. Portable saw mills were set up especially in Wynaad to turn out railway sleepers. In 1942, to meet increasing demands, a timber supply circle was introduced with powers to produce and supply timber from Madras, Travancore and Cochin. This circle was closed down in 1947. In June 1944, a retired conservator was appointed on special duty to survey and assess the damage to the state due to War. Besides the departmental fellings, a Kerala Timber Supply Corporation was formed in 1942 by prominent timber merchants of Malabar, which then supplied private timber for defence needs.

SECTION 2

Status of Forests of Kerala

Bamboo forests have generously fed three pulp and paper units at different points of time since the 1890s. They continue to provide raw material for more than 18,000 bamboo artisans. Bamboo brakes, as has been discussed in earlier chapters, is the result of retrogression of the forests over two centuries. It is necessary to have a brief look at the earlier status of the forests.

Early Accounts

Early accounts of the status of the forests of Kerala are recorded by Bourdillon (1892). He reports that the forests must have been more extensive, probably extending across the whole country. Vastly different forests are found in the north and south of the state. The low-lying country in the south is chiefly covered with tall grass and large trees scattered around. These lands are unsuitable for cultivation.

Kerala a small state, located in the south-west of India, has a land area of 38,863 sq. km. It lies between 8°18′ and 12°48′ latitude and 74°52′ and 77°22′ E longitude. The Western Ghats form the eastern boundary, and the Arabian Sea the western limit.

Relief

The Western Ghats, composed of hills, valleys and plateaus, stretch to a length of 600 km. Kerala has a coastline of 580 km (approx.). Geographically, the state is divided into the coastal, midland and highland zones. The highlands form 60 per cent of the total area. It is the zone which has the maximum forests. The midland has little or no forests.

Bourdillon speaks of the midland zones as bearing hill and scrub vegetation. Also, because of the moist soil conditions, this zone must have been covered with dense forests which were taken over for cultivation. Very old trees of the old forests were observed by Bourdillon still dotting the northern region.

He classified the vegetation as he saw it a century ago into four types: (a) Heavy moist forests of evergreen trees confined to the hill slopes of the north and to one-third of the upper hill plateau (b) Land originally covered with moist forests but now over-spread with scrubs of various ages (c) Deciduous forests (d) Rocky land covered with short grasses serving as pastures.

A Classical Classification is a Three-fold Division

The wet evergreen forests must have once been the most predominant. It has a large proportion of *Cedrela toona, Dysoxylum malbaricum* and *Diptocarpus indicus*. In the moist deciduous forests, *Terminalia* species, *Tectonia grandis*, *Dalbergia latifolia*, *Lagostromia lancelata* and *Xyla xylocarpa* are the main species. The third type is the sub-tropical broad-leaved montane forests, referred to as the *sholas*. Here the trees are stunted, while algae and lichens flourish. The predominant trees species are *Eurya japonica*, *Gardenia obtusa*, *Michalia nalgarcia*, *Ternstromia japonica* and various *Lauracea* species (Lal 1988: 54).

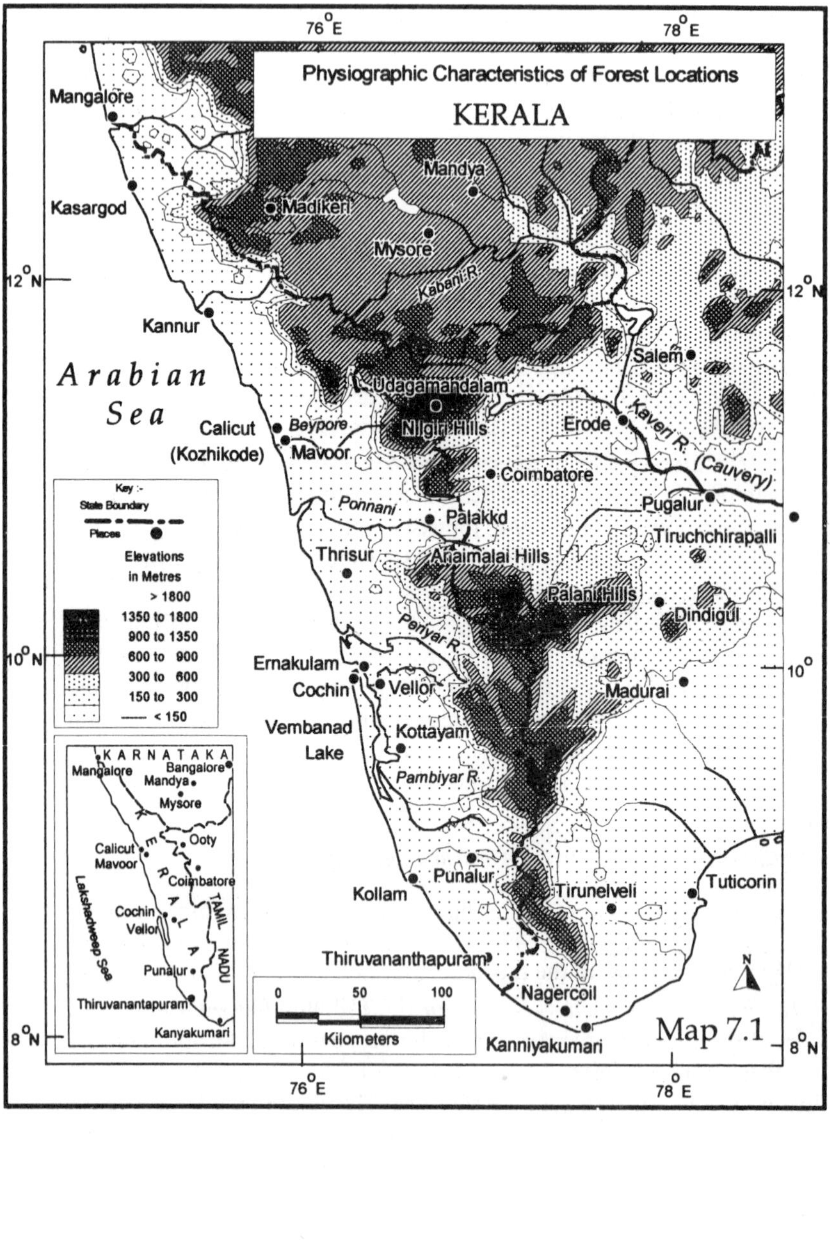

Physiographic Characteristics of Forest Locations
KERALA
76°E
78°E
12°N
10°N
8°N
Arabian Sea
Mangalore
Kasargod
Kannur
Calicut (Kozhikode)
Beypore
Mavoor
Madikeri
Mandya
Mysore
Kabani R.
Udagamandalam
Nilgiri Hills
Salem
Erode
Kaveri R. (Cauvery)
Coimbatore
Ponnani
Palakkd
Pugalur
Tiruchchirapalli
Thrisur
Anaimalai Hills
Palani Hills
Dindigul
Periyar R.
Ernakulam
Cochin
Vellor
Madurai
Vembanad Lake
Kottayam
Pambiyar R.
Punalur
Kollam
Tirunelveli
Tuticorin
Thiruvananthapuram
Nagercoil
Kanniyakumari
Key :-
State Boundary
Places
Elevations in Metres
> 1800
1350 to 1800
900 to 1350
600 to 900
300 to 600
150 to 300
< 150
KARNATAKA
Bangalore
Mandya
Mysore
Ooty
Calicut
Mavoor
Coimbatore
KERALA
TAMIL NADU
Lakshadweep Sea
Cochin
Vellor
Punalur
Thiruvanantapuram
Kanyakumari
0 50 100
Kilometers
N

Map 7.1

Grasslands on the mountains are not climax vegetation in the uplands of Kerala, they are a result of clearing of forests and failure of forests to re-colonize due to repeated fires (Lal 1988: 28).

Evergreen forests occur to a smaller extent in the southern districts of Trivandrum and Quilon. Towards the north, the whole of Malayattur reserve is more or less a continuous stretch of evergreen forests. Further north in Calicut and Malappuram districts the evergreen forests are found at elevations above 1,000 ft. Here the ghat slopes reach to elevations of up to 400 ft above sea level, bearing good evergreen forests.

The semi-evergreen forests are found adjoining the evergreen area and is in a transitional stage from the evergreen to the deciduous type.

Basha (1990: 132-43), reports that tropical moist deciduous forests constitute 4,010 sq. km, and form 42.66 per cent of the total forest area of the state. In many areas these forests are formed as a result of the disturbance to the evergreen and semi-evergreen forests as seen in the high rainfall (3,500 mm/yr.) areas of Palghat valley. These forests are found on the western and southern slopes up to an elevation of 250 m. Where soil moisture is impoverished and poor, they spread to an altitude of 600 m in the eastern slopes. In Attapaddy valley where shifting cultivation was practiced by the tribals, these forests are found at altitudes of 900 m also. The soil, temperature and rainfall that nurture the forests are given below.

Soil: In almost all the parts of Kerala, adjoining semi-evergreen patches, in the areas where the terrain is flat and the soil clayey, moist forests bearing teak are found, e.g. Thirunelli (Wynaad forests): at places like Thekkady and Idukki (highlands) the soil is loamy, deep and reddish in colour. Rainfall is between 1,500 mm per year. Such areas have a higher proportion of teak. Dry forests bearing teak are found in areas with rainfall below 1,500 mm per year. The soil is highly eroded and subjected to fire.

Rainfall: Kerala has a tropical climate of high rainfall and temperature. The vegetation is varied due to changes in climatic and altitudinal factors. The composition and extent of the forests have been drastically influenced by anthropogenic factors with the resuilt that major forest areas have vanished and only a small portion keeps its pristine character (Basha 1990). Bamboo could be either a natural undergrowth of the tropical forests or a result of their retrogression.

Distribution of Bamboo

During his travels through the forests of Travancore, Bourdillon (1892), observed *Ochlandra travancoria* forming the undergrowth of forests over immense areas in different parts of the country, and near the crest of hills it occupied the whole ground, covering the slopes with dense and almost impenetrable thickets. Incidentally, *Ochlandra spp.* belongs to the bamboo family and is referred to as reeds.

Ochlandra is widely distributed and species growing in abundance are *O. travancoria, O. travancoria* (var) *hirsutta O. scriptoria.* Also one species each of Bambusa and Dendrocalamus are in demand. Reeds are found in the evergreen and semi-evergreen patches also along the stream banks in the lower elevations (Basha 1982). Basha goes on to say that the reed areas have dwindled in extent, remaining only as discontinuous patches with poor growth.

Permanency of reed growth is brought about by impeded drainage or high rainfall. The reeds may grow as a pure crop or an undergrowth in the evergreen forests where the canopy is closed (Basha 1982). Two types of reed formations occur in Kerala, viz., the primary and secondary formations.

Primary Formations

Pure reed formations with neither undergrowth or overgrowth occur in certain localities like Pooyankutti, Gooderical, etc. (Basha 1982). These are stable climaxes perpetuated due to their peculiar flowering and copious regeneration. If the soil in adjacent areas is identical, the growth spreads and these are edaphic climaxes (Basha 1982). The primary formations of the reeds are found growing in association with certain tree species. Such patches are found in the Silent valley and are of two types.

Large stretches of reed associated with evergreen trees like *Calophyllum elatum*, *Bischofia javanica*, *Poeciloneuron indicum*, etc. In these patches stablity is maintained, since the closely growing reed does not allow regeneration and the high moisture prevents fire. These patches are found at higher elevations.

At middle elevations, reed grows as understorey in the evergreen forests. But here the growth is not thick enough to prevent regeneration of other tree species (Pascal 1988).

Secondary Formations

Secondary formatins of reed come up in abandoned shifting cultivation areas. Bourdillon (1892), in his accounts of the forests of Travancore, notes that the land cleared for cultivation was covered by 'heavy moist forests'. The species of reed which take over are *O. travancoria, O. scriptoria*, and, less often, *O. beddomeli*. This condition will lead to semi-green or evergreen with an undergrowth of *Ochlandra* (Pascal 1988).

Pascal (1988), considers the formations of reeds as *Ochlandra* brakes brought about due to anthropogenic interference. He relates the origin of these 'brakes' to two reasons: (a) of edaphic origin localized near water courses with poorly drained and hence a climax *facies* is formed rather than a degradational stage; (b) as a result of over-exploitation and clear felling of primary forests and abandoning of cardamom plantations in the past, was one of the important causes for the appearance of *Ochlandra* brakes (Pascal 1988). Bourdillon (1892) laments that 3,94,000 acres of virgin forests cleared for coffee plantations were abandoned in 1877 when the prices of coffee fell. These formations will stabilize as long as heavy biotic pressure is kept away.

Forests fires usually destroy the overground portion of the evergreen forests but leave the rhizome intact. Fire also cleans up the vegetation and opens out the canopy for the *Ochlandra* to spread (Pascal 1988).

Repeated yearly working of *Ochlandra* without regard for the closure season from June to September depletes the forests (Basha 1982). Clear felling and anthropogenic pressures in extreme cases not only alter the structure of the reed formations but also lead to the inversion of the grasses resulting in grasslands or Savannas (Pascal 1988, quoted in Basha 1990: 8).

It has been mentioned that there are other types of commercial bamboos in Kerala which both traditional and modern industrial sectors use. Like the reed, *Bambusa arundinacea* also occurs widely in evergreen forests as undergrowth and often as pure patches in moist, deciduous forests. In drier localities *D. strictus* grows in thickets (Basha 1988: 139-40). There is still another source for bamboo that is currently being exploited to the maximum. These are from the homesteads and farmlands where bamboo but not reed was cultivated, says C.N. Krishnankutty (1991: 1). On the basis of his study of homestead bamboo estimates, an average daily outturn of 0.93 million tonnes of *B. arundinacea* of which 57 per cent is consumed by the traditional sector and 27 per cent by the Kerala based PPI, while 13 per cent is exported. Other sources, in this case, the Mysore Paper Mill, claimed it that it bought homestead-grown bamboo from Kerala. C.N. Krishnankutty's estimates were carefully based on the actual ground space that was occupied by bamboo clumps.

It would not be out of place at this point, to mention that homestead bamboo is given away as a distress sale item to bamboo contractors who, naturally insensitive to felling rules, merely clear-fell the clump and auction it to the highest bidder. This is a negative consequence of the breakdown of the *Tarvad* system in 1967 where huge family land-holdings were cut into uneconomic holdings (personal communication, R. Shankar and K.K. Seethalakshmi, March 1993).

Post-Colonial Forestry—Kerala

The state of Kerala has a geographical area of 38,863 sq. km of which 24 per cent is forested. Out of this 24 per cent one-fourth is degraded, as noted by the National Remote Sensing Agency (NRSA, 1991).

The current status is given in Table 7.1.

Post-Colonial Impact on Forests

While the colonial powers had a major role in altering the face of the forests in Kerala, post-colonial pressures on the forests were not far

TABLE 7.1: CLASSIFICATION OF FOREST AREA OF KERALA, 1991 BY TYPE AND LEGAL STATUS

	Percentage	Area
Forest Type wise 100 sq. km effective Forest Area		
Tropical Wet Evergreen and Semi-evergreen	36.00	3,470
Tropical Miost Deciduous	43.00	4,100
Tropical Dry Deciduous	1.7	100
Montane Sub-tropical and Temperate Sholas	0.7	70
Grass land	1.43	130
Plantations (Man-made)	17.80	1,530
Total		9,400
By Legal Status		
Reserve Forests	83.00	9,341
Vested Forests	16.00	1,889
Total		11,230

Source: Government of Kerala, Forest Statistics, 1990.

behind. Among the fresh pressures, the one that is generally highlighted is the pressure of an expanding population and inadequate machinery to implement laws. Vishwanathan (1982) puts down 12 per cent loss of gross forest area from the 1950s to the 1980s. A close second is to meet the industrial demand for pulpwood species, 15.32 per cent of its natural forests, mainly moist deciduous, were converted to plantation (Basha 1990). The pressures created by the central government are deeply resented, those by PPI only surmised but not analysed, and those by UN FAO ignored, for unknown reasons. But several FAO officials mentioned that they were ignored and that Indian forest officials were un-cooperative (personal communication, October 1995).

Inroads were first gained into the forests in the wake of the 1946-7 Grow More Food campaign. This scheme metamorphosed into an onslaught on the state forests.

In the Malabar area the destruction was in the private forests, while the government forests were spared. The private forests covered a larger area and bore the brunt of land-hungry people and timber thieves. These two must be thought of as separate entities. While government forests were 1,400 sq. km, private forests were 3,000 sq. km. Land-hungry people, apart from local people, included migrants from Travancore and Cochin. The *jenmi* system of Malabar came up since the vast tracts of forest land, unoccupied at the time of land revenue settlement nearly a century ago, were considered as *jenmam* lands. In 1945, 3,100 sq. km belonged to 116 individuals. But at the time of the encroachments, the fact that a national asset was being lost irrespective of ownership, was forgotten.

Till World War II the private forests of Malabar were untouched. By 1945 the Government of Madras was concerned about the large-scale alienation of land by private forest owners. The Madras Prevention of Private Forests Act was issued in 1949 to prevent indiscriminate loss of private forests. But this act was insufficient, it only required that permission be taken from the Collector before any alteration was done or before trees were cut. This was intended as a temporary measure pending further legislation. It is perhaps here that the tea-chest maker who gained control over timber trees (chest-making was a lucrative business), and then with the coming in of FAO graduated into plywood and veneering which is dealt with a little later. This information is incidental to the question that is discussed.

In 1956 the State of Kerala was formed and the implementing of the Prevention of Private Forests Act became difficult. The owners of private forests were not interested in preserving their forests, only in easy money. The system of allowing cultivation only saw these *jenmis* permitting cultivation by different people every year; there was no legal alienation of the land.

In 1971 the Kerala Private Forests (Vesting and Assessment) Act was passed. The Kerala High Court struck it down. In 1973 the Supreme Court upheld the original Act. During this short battle, large-scale felling was carried out for timber. By the time the Act was enforced, nearly half the original area of private forests had come under cultivation. Thus the government took over only 1,900 sq. km of private forests.

Encroachment: Land-hunger

The densely-wooded Kerala had very little arable land for cultivation in its precious forests. Despite the immense land-hunger mentioned earlier, and the state and the FD seeing to it that no incoming family occupied more than 1 acre, the landless *did* encroach on uncultivated patches of land says Manoharan, the Conservator of Forests (personal communication, March 1993). These patches were left uncultivated due to high agricultural wage labour rates. All the foresters and environmentalists were unanimous that encroachment was really by the politically powerful thugs who established vast rubber and other plantations.[3] These encroachers were also accused of having abetted cultivation of tapioca on the hill slopes by their labour force; tapioca, a cheap substitute for grains, is the staple food of the labour force. But tapioca is a tuber and ecologically destructive when cultivated on the hill slopes. Again, in 1971 a central government team which visited Kerala asked the FD to allot land to the landless labourers 'under the arable land scheme' on condition that it would apply soil

Area in ha	Nos. of holder	Percentage
0.01–0.5	2,930.4	76.6
0.50–1.0	439.0	11.0
50 and above	0.4	Negative
Total area under tapioca	193,800 ha	production–32 lakh tonnes

conservation methods. The FD agreed under protest (see letter 2L11/71 dated16 June 1971, addressed to Department of Agriculture, New Delhi).

Government of Kerala (Forest Statistics 1990: 124), give the following break-up of operational holdings.

Meanwhile, in the Travancore Cochin area, the land-hungry people encroached upon the government forests. During 1946-7 several encroached areas inside the reserved forests were leased out for cultivation for a period of three years after which the land was to be returned to the Forest Department. In the hurry to implement these leases, proper surveys of the area were not made. As a result of this, people were actually taking over larger areas than those actually allotted. Leases were also given to organizations and cooperatives. These leases were given in the names of leaders. These leaders collected the rent and were supposed to hand it over to the department, which never happened. Amidst all this confusion and treachery, after three years the food shortage still remained, so the leases were extended by a year.

The many forest cases remained pending. The encroacher was to pay a fine per acre of area and was allowed to tend the crop till maturity and thereafter to harvest it. This amounted to nothing but regularizing the illicit occupation.

The government set up a committee in 1962, the Radhakrishna Menon Committee. The recommendations were that all pre-1960 encroachers should be regularized and the post-1960 ones evicted. New forest boundaries should be laid. People opposed the evictions when they were enforced, and another committee was set up in 1965. The Maniyangadan Committee recommended that no segregation of encroachers as pre-1960s or post-1960s be made. But the government did not agree, so the time was extended to 1968. Meanwhile several encroachers had filed writ petitions and got stay orders on evictions. The government further extended the year of eviction to 1977. This total vacillation on the part of the government machinery only assisted the encroachers to gain.

Nair (1985), points out that there was an absence of a clear-cut policy at the state level, so that the state government had to yield to local pressures from the agriculture sector. Social and political pressures have led to periodic regularization of encroachments, although this is a major offence under the Forests Act.

TABLE 7.2: AREA UNDER SELECTED PLANTATION CROP 1986-7

Crop	Area in '000 ha	Production in '000 tonnes
Rubber	347.8	202.12
Cardamom	62.7	2.50
Coffee	65.6	23.55
Tea 34.6	50.33	
Cashew	133.6	88.71

Source: Government of Kerala—Forest Statistics 1990-2.

Encroachment—Selected Forest Plantations

Apart from land-hunger, there was encroachment for the purpose of setting up of rubber and coffee plantations, coffee planters being a politically powerful group. Expansion of rubber plantation is due to the subsidy, offered by the State Government of Kerala, 1990. The negative aspect of plantation in forests is discussed in Chapter 6, Karnataka. Now tea is said to particularly injurious to the forest eco-system due to its tap roots. The once-green forests are now a mosaic of many hues ranging from pale green of the tea to deep purple of the rubber. The extent of area under plantation is given in Table 7.2. Timber and teak plantations are dealt with in a separate section. Tea and rubber are ecologically unfriendly, they draw in a lot of water.

The question of eucalyptus and teak as well as other timber plantations and their impact on the natural forest can be appreciated only after the question of the FAO is dealt with.

Why FAO

FAO was a US sponsored UN organization set up in 1945. The mandate of the FAO's forestry wing was to find pulpable tropical hardwoods to boost paper production both in the US and abroad, as pine, the pulpable raw material in US at this point of time, was still inaccessible (UNFAO 1954: 64) and Canada had not yet been drawn into the US shadow.

It was only in 1952 that the UN Secretary General sent out an invitation to member countries to join in joint explorations. But even before India formally agreed to this, FAO's forestry wing started its activity in this country, states the *Journal of Forestry* (vol. 49, no. 8, August 1951) and reports that agreements were made with India, Burma and Ceylon to provide technicians to instruct local persons in the best method of conducting logging, extraction operation and saw milling techniques. India, with its efficient forest service, was thought of as a useful base for FAO's operation in the south-east (FAO 1968: 9, 35, 47). FAO officials complained that Indian foresters were resentful and refused to cooperate. This is perhaps the reason why it finally chose Bangkok, Thailand being more pliant than India in the early 1950s.

Eucalyptus Plantations

In 1958, the communist Government of Kerala was dismissed by the centre and Birla found an entry. The Chief Conservator of Forests complained that 30,000 ha of forest land were converted to *Eucalyptus grandis* for the first time in 1958, on secondary grass lands on higher elevation above 2,000 ft (Govindan Kutty, personal communication, March 1993). These were in Peermed and Vaudiperyar. FAO in its pre-investment survey of Forest Resources in India (1970) had identified Chevvara and Piravana and noted with satisfaction that 24,000 ha of government plantations were raised and 250 private plantations covering an area of 4,140 ha in Chevvara were almost entirely of *E. grandis. E. grandis* was hit by the pink disease. Its growth was stunted. Plantations of *E. tereticornis* established at a lower altitude escapes the fungal attack. Kallarackhal plant physiologist at KFRI has made comparable studies of *E. grandis* and *tereticornis* and found that the latter does not close its stomata and therefore its effect on the ecosystem is more damaging. This issue has been discussed in Chapter 6 under MPM Plantations. FAO (1970) states that by 1967, 24,000 ha of government plantations of *E. grandis* are at two sites. Piravana and Chevvara and there are 250 private plantations covering an area of 4,140 ha of *E. grandis* at Chevvara.

Foresters by and large abhor imposition of eucalyptus by the central government; its plan grants are conditional and earmarked for eucalyptus. Note for instance in a letter no. 21-11/71 by the Kerala Forest Department to the Ministry of Agriculture, Government of India; a report is presented on the action taken to implement the recommendations of the central team which visited the state in 1971. The team recommends that the tempo of raising industrial plantations must be kept up otherwise it would jeopardize the newsprint mill in the state. Action was being taken to achieve the target of 8,000 ha of plantations of this fast-growing species. Another valuable suggestion which the team makes is to allot lands to the landless labourers under the arable land scheme but with the condition that the cultivator will adopt soil conservation methods. The Forest Department (FD) states that it agreed to set apart certain forest areas under the arable land scheme but under protest.

The forestry activities in the state during the fourth five-year plan (1969 -74) were targeted at raising 7,000 ha of teak, 500 ha of softwood, 8,000 ha of fast growing species, 500 ha of fuel plantations and 500 ha of miscellaneous plantation, bamboo, *balsa*, *anjili*.

Social Forestry (SF)

Natural forests provide goods and services to the local people with colonial control over forests and forest produce, and much more so after certain industries became the main beneficiaries of forests, the local people were over a period of time deprived of all their right to the forest and to its products. Social Forestry was introduced in India to placate the growing unrest of these deprived people and funded by international organizations like the World Bank, SIDA, etc.

In Kerala however, this is not exactly what happened. Rural Kerala is dependent on coconut shell and not on forest faggots (Thampi 1983: 121). There is no information on other altenate goods and services to those provided by forests. It appears that SF is a statutory necessity and was introduced as late as 1984. Local species were generally raised, the first crop had not yet been harvested, and the CCF was not clear what use the felled timber would be, the benefits of the sale would be shared in the proportion of 75 per cent to the government and 25 per cent to the local people. Social funds were diverted to general forestry; *Acacia auriculiformis* was perhaps imposed on Kerala foresters too. As a part of Social Forestry this exotic species lines the roads four deep as avenue trees.

Kerala Forest Research Institute scientist Shankaran has pointed out the negative feature of its leaf litter on the soil. This is discussed in detail in the Chapter on Karnataka under MPM forest plantation.

An interesting observation was made by the conservator, the Working Plan Division: A third of the eucalyptus plantations that were felled in the Peermed grasslands has reverted to natural forests (Meher Singh, personal communication, March 1994). Earlier it was mentioned that grasslands were not climax vegetation but a result of repeated fires. Singh also mentions that plantations are on a decline.

The pressures on the Forest Department to yield more and more forest as lease area by the Birlas is discussed under the head 'Entry of the Modem Sector'. The growing need is not due to increase in production, it cannot be so, as bamboo if treated with respect, will continue to yield. The careless felling by the agents of HNL as well as the Bamboo Corporation is also discussed in the section on 'Bamboo and Reed Felling Practices by Traditional and Industrial Sector'.

These being the observations of the team and its understanding on the forestry issues in Kerala, it would be worthwhile mentioning some viewpoints gathered from a leading local daily. The current situation in Kerala is that, more and more land is being given to the industries to raise captive plantations. 700 ha land were given to HNL this year to raise eucalyptus (*The Times of India*, 6 June 1994). This, when the Forest Research Institute (FRI), Dehradun, has come out with a publication on eucalyptus which casts gentle aspersions on the merits of this tree. This continuing handing over of land to industries for raising the disputed species is under the directive of the centre (*TOI*). There are many more controversies, with regard to forests coming up in Kerala. Continued allotment of land (especially in a land that has so little wasteland) for industries needs to be questioned. It is interesting to note that leading personalities like Satish Chandran Nair (Chairman, INTACH) states that 'the Forest Department in its present set-up cannot be trusted to take care of the forests' (*TOI*).

Kerala Forest Department: Forest Circles

For the purpose of administration, Kerala Forests are divided into five circles (a) Southern Circle with Quilon as headquarters (b) High Range—Kottayam (c) Central Circle—Trichur (d) Olvakot Circle with Calicut as headquarters (e) Northern Circle—Calicut and Palghat.

The Kerala Forest Department promised, but failed to give the actual annual division/circle-wise outturn of bamboo and eucalyptus. It provides only broad details which are given in Table 7.3. It is not clear why Grasim is provided immense quantities of bamboo when its production in the Kerala plant ceased in 1983. The Grasim is Aditya Birla's flag ship and is doing well. It is not clear whether the raw material was shipped out to its plant in Gujarat and in Kalyan (Maharashtra).

Kerala Forest Development Corporation (KFDC) is also expected to provide raw material to the Pulp and Paper Industry as well as to Veneer and Timber Industry. Its data is given in Table 7.4.

TABLE 7.3: KERALA FOREST DEPARTMENT'S ANNUAL COMMITMENT TO VARIOUS PPI UNITS

Year of agreement	Type of wood	Quantity (mt)	Royaty rate (in addition to taxes)
1. HNL			
10 July 1974 for 20 years	*Eu. grandis*	1,00,000	Rs. 325 per mt
	Eu. hybrida	50,000	Rs. 325 per mt
	Reeds	1,89,000	Rs. 12 per mt
2. Grasim Industries			
27 Oct. 1988 for 5 years	Eucalyptus without bark	1,60,000	Rs. 250 per mt
	Bamboos	40,000	Rs. 250 per mt
3. Punalur Paper Mills*			
20 April 1982 for 10 years	Eucalyptus with	40,000	n.a.
	bark Reeds	25,000	n.a.

Source: PCCF's office (Letter no. G38191/93 dated 23 February 1993).
Note: *PPM stopped production in the early 1980s and closed in 1986.

TABLE 7.4: SUPPLY OF EUCALYPTUS BY KFDC TO VARIOUS INDUSTRIES IN KERALA

Year	Grasim Industries Ltd.	Western India Plywood Ltd.	HNL
1989-90	17,425.435	—	—
1990-1	23,356.527	10,143.210	—
1991-2	49,125.367	8,555.750	—
1992-3	60,57.085	26,462.148	371.553

Source: FFDC Letter no. G3571/93 dated 10 May 1993.

It is not clear why Grasim which officially closed in 1984 is supplied with eucalyptus. It is not clear what Western India Plywood does with eucalyptus.

Areas Reserved for HNL

Eucalyptus grandis plantations are located in Pamba, Kottayam, Punalur and Trivandrum Forest Divisions. With the closure of Grasim and Punalur since the last decade, all the plantations of *Eucalyptus tereticornis* were to be reserved for HNL. The government also agreed to raise

TABLE 7.5: ROYALTY RATE FOR HNL

		Royalty rate
Eucalyptus Grandis	1,00,000 mt	Rs. 335/mt
E. Tereticornis	50,000 mt	
Reeds	1,89,000 mt	Rs. 12/mt

Source: CCF's office letter dated 26 June 1993, no. G38191/93, Agreement on 7 October 1974 - HNL.

more plantations to tide over any future shortage or to supply more in case of an increase in the production capacity of the mill.

In the case of reeds, six felling series were earmarked for HNL, viz., Vazhachal, Edamalayar, Pooyamkutty, Adimally and Chalakkayam, the royalty rate being Rs. 12 per tonne (at 50 per cent moisture) which was initially valid for 20 years but appears to be unchanged even today according to the letter from CCF office, no. G38191/93 (refer to Table 7.5).

The royalty rate for eucalyptus was fixed at Rs. 11 per tonne of green wood (50 per cent moisture) with an option to revise after five years, the increase to be proportionate to the rise in the price of the product of the company. Today, according to CCF office letter, no. G38191/93 dated 26 September 1993, the royalty rate is Rs. 335 per tonne.

Reed was supplied at the royalty rate of Rs. 12 per tonne (at 50 per cent moisture) which was valid for a period of twenty years (refer to Table 7.6).

One of the articles (Anon, 1983) in the *Financial Express*, while complaining of wastage of precious raw material pointed out that the extraction of raw materials had started in 1978 when the plant was still under construction. It also noted the faulty way the eucalyptus and reeds were collected and stocked when the mechanical pulping unit was in a working condition. Estimates of the loss reveal that it was just a little less than Rs. 5 crore in 1981-2. The pulping value of the debarked wood eroded resulting in an estimated loss of Rs. 50 lakh.

TABLE 7.6: RAW MATERIALS PROVIDED BY KFD (1992-3)

Bamboos	57,21,455 (mt)	
Eucalyptus tereticornis	855.655 (mt)	(50% moisture content)
Eucalyptus grandis	38,963.179 (mt)	(50% moisture content)
Reeds	54,223.192 (mt)	

Source: CCF's office letter dated 15 June 1993, no. G38191/93.

The loss to the forest department is even higher since it gives wood at a subsidized rate of only Rs. 12 per tonne for eucalyptus and Rs. 11 per tonne for reeds. In terms of open market rates the loss could therefore exceed Rs. 1 crore. The *Financial Express* correspondent, however, fails to note that the HNL plant was expected to go into production in 1978 but could not, due to teething problems. This has been the fate of almost every unit in the private sector, and to highlight a public sector alone as an example of irresponsibility and poor planning shows poor judgement.

Observations on the Felling of Reeds

Ochlandra travancoria and *Ochlandra rheedii* were the two common species of reed-like bamboo to be found in Kerala. *O. rheedii* is shrubby in nature and grows in thick clumps along the river banks in wet evergreen and semi-evergreen forests. HNL's forest officer stated that young shoots of reeds grow very rapidly and attain full height and thickness within a month of sprouting. A visit to Vazhachal, 85 km away from HNL, brought to light that the agencies entrusted with its extraction flouted all basic rules of felling (for details see section on forests).

FAO and Status of Kerala Timber Forest

Earlier we had mentioned that FAO's mandate went beyond managing pulpwood to develop PPI. It also went beyond agriculture. FAO rationalized this and tried to show a theoretical link between forestry and agriculture, that is, the forests' role in improving hydrology and in soil protection apart from providing fuelwood for the village folk. It emphasized the importance of forest-based industry, such as saw mills, plywood, veneer, particle boards, pulp and paper, to improve the living standards of the people. It stated that FAO's Industries Division is working towards this end (FAO 1968: 9, 35, 40) especially since the rich and varied forest resources are a good source for export; and since neither the US nor Europe has timber comparable to the tropics. FAO's main argument was, and perhaps still is, that these products will improve the balance of payment which is imperative for those who cannot produce sufficient food for themselves (FAO 1968: 54, 57, 59) nor control its growing population.

In its Pre-Investment Survey on Forest Resources—India (1970), it suggested more intensive forest management. Kerala is the only state

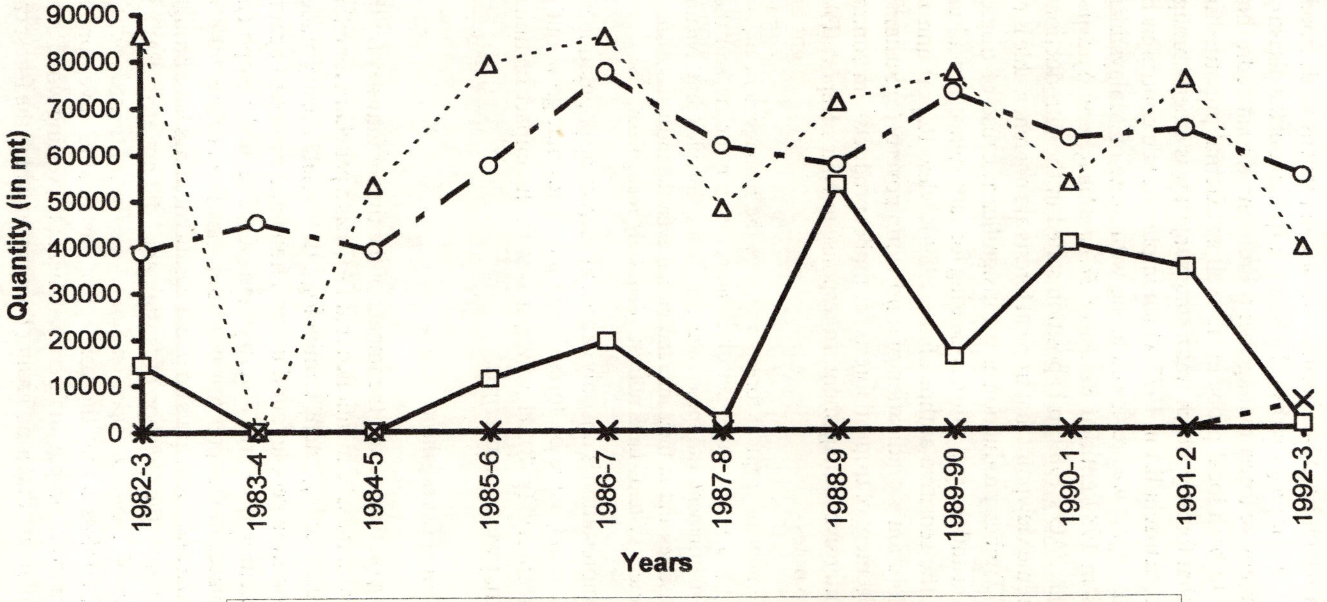

Source: WCPM, September 1993

Figure 7.1: Quantity of Forest Raw Material allotted to HNL

in India on which FAO has brought out a comprehensive document (FAO 1985). It was prepared by CTS Nair who points out the disastrous effects of intensive forest management. From the Kerala cadre, Nair is sensitive to both local problems as well as external pressures and points out that forestry which was a state subject was quietly brought under the concurrent list in 1977 which implied subtle control by the Centre. This report was reviewed along with those of Malaysia and Philippines in 1989 entitled *Review of Forest Management Systems of Tropical Asia*. FAO's 1989 (p.1) position stated in the introduction is (a) The extreme viewpoint that tropical forests are not renewable is not correct (b) Timber production is an insignificant cause of tropical deforestation and the real cause is clearing for subsistence agriculture. A.J. Leslie the sensitive author of the 1989 Review Report and his suggested solution is eradication of third world poverty (c) Sustained yield management of tropical forests for timber products—in contrast to earlier intensive management. In response to Nair's Report (FAO 1984) Leslie notes

> that management of tropical forests involves the management of, and adjustment to, a much wider set of elements of the forests and their immediate environment and these elements are too complex and dynamic both in themselves and in their interaction that even the most successful of the management systems have not yet adequately recognized.

He adds, 'deforestation is not only caused by increasing population but by rising demand made by forest based industries that grew out of economic recovery' (p. 8). However, it is with the impact of timber industries that we are concerned.

Sustainable Forest Development

Before going on to evaluate the impact of the timber industry on the Kerala forests it must be mentioned that FAO has a penchant for either coining or resurrecting catchy terminology. Its 'Social Forestry' ended when the common people found it was a fraud and protested against it. A combination of 'Farm Forestry' and 'sustainable development' took SF's place. In Chapter 5 it has been shown that farm forestry with casuarina and eucalyptus on poor peasants holdings have brought no development—the question of 'sustainable' did not arise. FAO, in its collection of *Technical Papers on Sustainable Forest Management*, FP 122 (1994), confesses that the concept of sustainable development (SD) is deceptively simple; in forestry, it meant harvesting wood at an average annual rate that is no greater than that at which the forest can grow. Apart from the fact that such measurements are difficult to make, humid tropical forests are complex and that complexity itself is the first limiting factor (1994: 20). Leslie proceeds to burst the bubble:

> Apart from technical difficulties—which are not the major limiting factor the 'present obstacles are much more the result of malfunctioning, lopsided inequitable world economy. Unless a way can be found to escape from the economic stranglehold on tropical forestry management of sustained yield wood production is a barely realistic target, let alone sustainability in all its facets.' (FAO 1994: 25)

Intensive Forest Management (ISF) in Kerala

Selection felling based on working plans was the rule in Kerala. ISF was introduced by FAO—so too processing of wood for ply and veneer. Roads were opened up to reach the inner-most forest preserve to get at the natural timber especially teak and rose wood that abound in Kerala forests. Kadar Kutty was the only person who made tea chests for the increasing tea trade in the western ghats across the three southern states in 1944. He was given permission to extract timber in 1946 and Forest Industries Travancore Ltd., was set up in Alwaye. With technical help given by FAO and the unit set up by GOI, again with FAO's help at Aranya Bhavan, Bangalore, Kadar Kutty graduated into a monopoly holder of plywood and timber industry of the region. Official records point out that Western India Plywoods was set up in 1945 near Cannanore (Malabar) and later another unit was set up in Dandeli in Karnataka. It earned the name of being the largest manufacturer of hardboards and plywood. He has been exporting decorative veneer and hardboard to USA, Europe, UK, Sweden, Japan, Holland, Hongkong, Africa, Saudi Arabia, Iran, Iraq, Kuwait and eastern countries (*Economic Times*, 11 January 1984).

The only other company in Kerala is Veneers and Laminations India Ltd., incorporated in 1974, a small joint sector unit with J.K. Jacob at Tiruvella. FAO notes that Kerala and Karnataka account for 19.5 per cent of installed capacity of plywood (FAO 1989: 40). The north-east tops this list.

Kerala Forest Conservators and Forest Economists Reaction

The negative effects of the forest-based industry on the forests of Kerala are perhaps best reported by a sensitive forester and a forest economist.

CTS Nair (FAO, 1985: 73), points out, that technically, harvesting of timber on a sustained basis from a natural stand requires that the yield be regulated to prevent over-cutting. Conceptually, harvesting is linked to the idea of rotational age but trees are harvested as they become mature. Increasing economic pressure for supply of timber has frequently caused a downward revision of both the area under protection as well as downward revision of girth limits, for felling of the commercially useful species.

It is not surprising that a forest economist, Mammen Chundanamannil (1986: 112-17), found that 80 per cent of FD's investment was on extraction of timber which yielded 80 per cent forest revenue. He also found that while the timber yield increased five-fold, the expenditure per unit of production declined. Investment was not consistent with sustained yield. Long-term investment was concentrated on two species, eucalyptus and teak.

All this despite the Tree Protection Act passed in 1974, two years before Karnataka did so. Opinions are divided among foresters—Vishwanath in 1982, complained that 12 per cent of the forest cover is lost within thirty years 1950 -80. Pasha is angry that the FD panders to the industry's need. But CCF Govindan Kutty says all clear felling ceased in 1982 and selection felling is not followed due to lack of stock (personal communication, March 1993). Singh concurred that government had decided to stop working old areas which were over felled while Kerala FD has stopped converting natural forests into plantations of either *Eucalyptus grandis* or *E. tereticornis* (personal communication). NABARD's Director of Forestry, Haque, was surprised that Kerala is the only state which had not asked for funds for plantation activity, and HNL withdrew its request (Haque, personal communication, July 1995).

SECTION 3

Hindustan Newsprint Limited (HNL)

No Private Investment in Newsprint

The paper industry has made considerable progress in India since the installation of the first paper mill with the Fourdrinaire type machine in 1832. After 1947, with the help of the government, the newly-emerging industrial houses set up pulp and paper industries. But none came forth to produce newsprint as investment was heavier, returns lower, the gestation period longer and there was no precedent in the production of newsprint in India.

In the first two plans, i.e. 1951-61 and 1956-7, the paper industry made satisfactory progress with an average growth rate of 15 per cent and 10 per cent per annum respectively which was achieved through establishment of new units as well as by 'modernization' and more correctly through renovation and expansion of existing mills. By 1961, there were 28 units in operation with an annual installed capacity of 4,10,000 tonnes and annual production of 3,05,000 tonnes. But there was still no newsprint. It was only in 1956 that the first newsprint mill (NEPA), was set up as a public sector unit (see NEPA in the section under Madhya Pradesh—Ch. 3).

Newsprint in the Public Sector

Till 1955 the entire requirement of newsprint in the country had to be met through imports. Indigenous production of newsprint commenced with the starting of the first paper machine at NEPA in 1955. Subsequently the second machine was added in 1969, and started production. But the government's policy of control through its two ministries and imposition of wrong untested raw material made it impossible for NEPA to produce good quality newsprint. Besides, there was pressure from the newspaper lobby to import more newsprint. The import of newsprint during 1969-70 was about 1,50,000 tonnes as against the indigenous production of 30,000 tonnes. The demand was inflated by dealers who put in false claims for non-existent newspapers—'benami' as it is popularly known. The demands were through the Registrar of Newspaper Industry to the Ministry of Information and Broadcasting (I and B). The purpose of the agents was to corner and sell the imported newsprint at inflated rates whenever there was shortage. The demand for newsprint during the year 1976-7 was estimated at 2.50 lakh tonnes and was expected to go up to 3.50 lakh tonnes by 1978-9, and to increase steadily thereafter. But this was still inadequate for the growing newspaper industry.

Newsprint must be cheap and also have basic strength characteristics to enable it to run on high-speed rotary printing machines used for printing of newspapers. The existing indigenous mill NEPA, was compelled to use *Salai* and later bamboo and eucalyptus as raw material. The new newsprint mill was meant to be based on hardwood. Pulping technology to utilize hardwood had to be updated by either importing or developing advanced technology of paper making. A joint FAO–Indian team undertook this task at the request of the GOI in 1970. Moreover, newsprint produced should cost less than writing and printing paper. Since import of technology meant heavy investment and because the price of newsprint was not attractive while return on investment was low, no private entrepreneur in India even in the 1970s came forward to make new investments. There was a need for increase in indigenous capacity, which apart from preventing a drain on foreign exchange, would also reduce India's dependence on the vagaries of the international market, from both fluctuations in price and on the supply front.

HPC/HNL as an Outcome of 'Scarcity'

The Government of India felt the need to go in for fresh capacity generation in the public sector in order to offset the scarcity likely both for cultural varieties of paper and newsprint in the country and set up the Hindustan Paper Corporation (HPC) and its subsidiary Hindustan Newsprint Limited (HNL) in Kerala. Our concern here is with the HNL.

To achieve these two objectives, HPC was asked to do several things (see our report on 'Why HPC'). The Kerala plant of the HPC was specially meant to produce newsprint and is referred to as HNL. At this point, it probably asked FAO for suggestions. Among FAO's several suggestions for sites in its *Pre-investment Survey Report* (1970), it specifically mentioned not only Kottayam but also Piravana or Chevvara where *Eucalyptus grandis* plantations were raised on its advice as early as in 1970. Not infra-structure, but raw material is FAO's only concern for choice of site. The Report points out that *E. grandis* yields good cold soda pulp for manufacture of newsprint besides rayon grade pulp of promising quality.

The objective of HNL newsprint project was to manufacture 80,000 tonnes of the newsprint per annum in Kerala. The project was cleared from the investment angle in August 1974 at an approved capital outlay of Rs. 82.88 crore. It was expected to go on stream by September 1978. The mill was designed to manufacture newsprint using a combination of Chemi-Mechanical Pulp (CMP) produced from eucalyptus wood and chemical pulp made from reed and bamboo in the proportion 70:30 in the furnish. This possibly was the consequence of the beginnings of one of the findings of FAO/GOI's team that was under the project management of UNDP from 1977 onwards. It suggested separate refiners for *bamboo* and *E. tereticornis* (1981: 22) the latter also being a part of HNL's newsprint furnish. It found Eta reed a good substitute for long fibre pulp, provided the other furnish component contributed adequately to bonding (1981: 25). *E. tereticornis,* however, does not do this; high black liquor viscosity was observed which could be reduced by increasing the free alkali in the liquor. Another disadvantage of the hardwood was poor evaporation and combustion of mixed hardwood-bamboo liquor. UNDP's contribution was $ 20,64,256 and GOI's Rs. 31.39 million (1981: 2).

HNL's newsprint was expected not only to save an annual foreign exchange amount of Rs. 30 crore, but also to further the growth of the new hardwoods technology which was then regarded as a major advancement over the use of native bamboo (Sarkar 1977).

According to B.P. Maurya, the Union Minister of State for industry in 1976, West Germany had provided DM 30 million (Rs. 10 crore) as a long-term loan for the purchase of a paper machine from that country, while the mechanical pulp plant would come from Sweden. But the chemical pulp plant and boiler equipment would be indigenously fabricated. However, there were delays due to non-availability of foreign machinery and equipment (*Financial Express*, 31 August 1985).

HNL's Teething Troubles

There was much speculation and a great deal of publicity when commissioning of the plant was postponed indefinitely.

Informed sources sense the invisible hand of some powerful private industrialists behind the delay in commissioning this prestigious public sector newsprint unit. They did not want the Vellor mill to stabilize itself for obvious reasons. With this end in view, these business tycoons had been manipulating things in the mill where they had successfully planted their henchmen in key positions (*Financial Express*, 17 October 1983).

Dalmia of Punalur Paper Mill of Kerala, for instance, had earlier approached HPC headquarters in Calcutta with a project to supply pulp to the Vellor project. Also there was the even more powerful Birla in Kerala, both being dependent on eucalyptus especially *grandis* and bamboo as the raw material for their paper and rayon production. They feared that they would face restrictions with the Vellor mill commencing production and consuming a large chunk of these raw material resources in the state.

HNL had been quoted as a classic profile of a public sector project in the doldrums, due to militant trade unionism in the state. It was also used as a convincing excuse for the reluctance of the Centre to invest in Kerala which had earned a 'bad' reputation not only for the activities of its labour but also for being left in its politics. This widespread temptation to blame labour for all the ills of the project is obviously unscientific as well as irrational. Workers had not asked for a wage revision for ten years from 1973 to 1983. Even when there was absolutely no labour unrest, the management was unable to hasten the work. A trade union source claimed that it had offered to give in writing to the management that there would be no strike until commissioning, if the date of commissioning could be announced. But the offer was turned down (*Financial Express*, 17 October 1983).

It is alleged that workers had been alienated to such an extent that there was virtually a communication gap between them and the management. The former learnt that the first newsprint sheet had rolled out of the mill at midnight on 27 February 1982, only when they actually saw the newspaper report next morning. It is reported that when the trade unions complained, the management's reply was 'after all the father is not permitted to see the delivery of the baby'. But the trade unions were kept in the dark about the Vellor unit of HPC being converted into its subsidiary and re-christened HNL in place of its earlier name—the Kerala Newsprint Project. HNL had its share of strikes but the *Financial Express* correspondent reports that most of these could have been avoided had there been efficient personnel management. He also thinks that many of the strikes were manipulated by the management. Last but not least, the multiplicity of labour unions had been the bane of the Vellor Newsprint Mill (*Financial Express*, 7 October 1983).

With regard to finance, there were other allegations. It is reported that the company had no financial discipline and budgetary control. Due to these reasons, an internal auditor had left the company in utter disgust, because it had neither a procurement policy nor a purchase policy (*Financial Express*, 17 October 1983). At a later stage the technical problems and reasons for escalation of costs will be discussed.

It is alleged that the HPC authorities for reasons best known to themselves had not been actively pursuing the project. All this naturally led to numerous delays and the newsprint project that was conceived in 1973 finally started commercial production only in November 1982, with an escalated cost of Rs. 157.802 crore (*Financial Express*, 31 August 1981).

HNL Takes Off

The company's records indicate that the mill went into production in 1982, and till November 1992 produced over 7.7 lakh tonnes of newsprint, saving the country over Rs. 620 crores in foreign exchange (HNL, March 1993). It certainly had its share of problems in the first few years. The main reasons for shortfall in production in the first three years were power shortages and non-stabilizing of the chemi-mechanical pulping plant (CMP). During the year 1985-6, however, HNL produced 78,634 tonnes of saleable newsprint, representing 98 per cent capacity utilization as against 80 per cent during the previous two years. The CMP plant stabilized and there was marked improvement in the production of chemi-mechanical pulp. However, a larger quantity of imported chemical pulp had to be used in the furnish on account of shortage of reeds. The target of production for 1986-7 was fixed at 80,000 tonnes, representing 100 per cent capacity utilization, but production was only 68,557 tonnes mainly due to power shortage (Rao 1989: 25).

HNL achieved an all-time record production of 90,388 mt of newsprint during the year 1990-1, attaining record capacity utilization of 113 per cent, as against the earlier record production of 86,844 mt with a capacity utilization of 108.6 per cent, achieved in 1989-90. The high level of production was possible because of investment in a most modern and highly automated newsprint machine manufactured by J.M. Voith of West Germany. It was also claimed to be the only newsprint machine equipped with thermocompressor type steam and condensate system. According to the mill's sources, the start-up and subsequent running of the machine had been very smooth and trouble free (*Economic Times*, 13 May 1983).

HNL was HPCs flagship mill. It maintained an extremely high profile during a three-year period (1986-7 to 1988-9) and its production was maintained between 86 to 102 per cent of capacity. The year-wise break-up of production was 68,557 tonnes (85.70 per cent) in 1986-7; 81,513 tonnes (101.70 per cent) in 1987-8 and 78,637 tonnes (98.3 per cent) in 1988-9.

It is worth noting that HNL made a record profit of Rs. 22.08 crore in the year 1988-9, thus helping in the recovery of the loss-making public sector unit. HPC considered HNL 'a jewel in its Crown' (*Financial Express*, 9 November 1989).

Its surplus further emboldened HPC to go in for a capacity hike of its unit to over 1 lakh at a cost of Rs. 150 crore without seeking any financial help from the Government of India, contradictory to its past policy (*Business Standard*, 5 December 1989).

In 1991-2 HNL surpassed all earlier records by achieving a production of 90,181 mt with a capacity utilization of 119.6 per cent. Since 1989, HNL has switched over to production of 49 gsm newsprint. This product, says a *Times of India* official in Bombay, is almost as good as imported newsprint. HNL's major investments have been on automation and upgradation of paper machines.

Newsprint production of the mill since commissioning is given in Table 7.7.

In addition to these, the reasons for a high production level as claimed by the mill, were continuous improvement in the efficiency of the various sections of the plant, proper planning and scheduling of maintenance, timely procurement of raw material, good liaison with government authorities and maintaining high productivity level with cordial industrial relationship. Based on its performance, HNL received the productivity award from the National Productivity Council for the years 1989-90 and 1991-2.

TABLE 7.7: PRODUCTION AND CAPACITY UTILIZATION BY HNL

Year	Newsprint production mt	Capacity utilization (%)	Net profit (+)/ loss (–) Rs. Million*
1982-3	25,196	31.24	–46.20
1983-4	64,176	80.20	–107.30
1984-5	64,217	80.30	–86.70
1985-6	78,635	98.30	–17.10
1986-7	68,557	85.70	–4.93
1987-8	81,513	101.90	
1988-9	78,660	98.30	
1989-90	86,844	108.60	
1990-1	90,388	113.00	+31.20
1991-2	90,181	119.00	
1992 till Dec. 92	71,225	126.00	

Source: HNL.
Note: *From Y.A. Rao.

Why Losses

It is important to explain the reasons for the losses in the first few years. This is mainly because the governmert chose to peg the price of newsprint (NP) to much below the cost of production. It worked out the cost of production on normative efficiency parameters to Rs. 7,000 per tonne. It failed to note the high capital cost, Rs. 1,560 million, when the trial production began in 1982, and that it was not Rs. 820 as was anticipated in 1978. It also failed to note several other factors such as that HNL uses unsuitable raw material—eucalyptus, along with reed and bamboo—in whatever proportion that is made available, and therefore its pulping process had to be adapted from the simple thermo-mechanical process used in the west for softwood to non-conventional chemimechanical to cope with addition of hardwoods (FAO 1981: 14). The modified process involves a much higher consumption of chemicals than anticipated. Again, even after commissioning the pulp plant, HNL had to use imported pulp due to power shortage. Power provided by the state was outside its control. There was a need to conserve energy on pulping, this being the most power-intensive operation. But again price was worked out on operating costs and normative efficiency parameters and revised to Rs. 8,000 per tonne in April 1984. All four newsprint mills asked for revision of prices till finally the Bureau of Costs and Prices relented; it fixed the price of HNL's 52 grammage newsprint at Rs. 8,960 per tonne. It is important to remember that Indian units lack economy of scale, apart from the reasons which explain the high cost of newsprint in comparison with the newsprint produced in the west. In 1988 the price was revised again, this time to Rs. 10,160 per tonne for 49 grammage newsprint (Rao 1989).

There are other reasons why HNL was making smaller profits, but the cost of production varies from firm to firm over time. Raw materials cost 30-5 per cent of the total cost of production, and chemical utilities conversion, 40-45 per cent. The cost of production by HNL is higher than that of Tamil Nadu Newsprint Limited (TNPL), because HNL uses reed and eucalyptus and the price of eucalyptus is rising, whereas the price of bagasse used by TNPL is almost steady. HNL had to use expensive imported pulp in 1984 and 1985 when it got no raw material from the Forest Department. Again, chemicals, utilities and conversion costs for HNL in 1988-9 was 60.3 per cent of the total cost while it was only 30 per cent for TNPL. In 1990-1 it ranged from 64.7 per cent for HNL and 39.5 per cent for TNPL, the bagasse-based unit (*Journal of Industry and Commerce*, 1993: 2-15).

Technology—Pulping

The raw materials used are *E. tereticornis* or a combination of *E. grandis,* reeds and bamboo. Debarked eucalyptus logs and reeds and bamboo are chipped. These chips are then transported to concrete silos and from there to the pulp plant through belt conveyors. Wood chips fed into the Chemi-Mechanical Pulp (CMP) plant are washed, steamed and impregnated with caustic soda. They are then mechanically pulped in three stages in raffinators. CMP technology yields as much as 85 per cent cellulosic raw material from bamboo, reeds and eucalyptus. TMP would yield 95 per cent. TNPL uses an even more advanced chemi-thermo-mechanical pulp. But its power consumption is very high and the yield comparatively low (*Journal of Industry and Commerce,* May 1993: 10). Also, HNL used the maximum amount of mechanical pulp to the extent of 75-80 per cent and is least dependent on imported mechanical pulp which was about 13 per cent in 1990-1 when the overall consumption of pulp was only 0.905 per tonne of newsprint. This has marginally increased to 0.910 per tonne after the change-over to 49 gsm (*Journal of Industry and Commerce*: 11). After going through the first raffinator the spent liquor is removed, and after the second stage of refining the pulp is bleached by hydrogen peroxide. One of the major problems faced by the industry in the manufacturing process is the accumulation of silica. The black liquor which is sent to the recovery plant precipitates lignin and silica (it may be noted here that while lignin precipitates at a pH of 10, silica precipitates at 10.1). The lignin (mixed with the silica) is used for burning and thus the digesters get coated with silica. HNL is the first plant in India to set up a desilication plant. This plant was designed by the Central Pulp and Paper Research Institute (CPPRI)), Saharanpur with assistance from SIDA and UNDO.

New Bleaching Technology

It is worth noting that HNL is the only pulp paper and newsprint plant in the country which uses and has pioneered the usage of eco-friendly hydrogen peroxide in place of calcium hypochlorite for bleaching the CMP, this being a part of its technological upgradation programme. It is well known that hydrogen peroxide is a very expensive but nonpolluting bleaching agent available only from one source and therefore can be starved of it. Hydrogen peroxide has another advantage over the conventional process of bleaching. The old calcium hypochlorite used to leave about 1.6 per cent of ash in the final product which tends to damage the paper machine. Hydrogen peroxide also means saving on time and labour. It also means an increase in production. If conventional pulp is used, the paper machine has to be stopped for an hour every day to clean the ash.

Secondly, with the use of hydrogen peroxide cellulosic content of the raw material that is utilized is as much as 85-90 per cent. It was also possible for production of newsprint with lower grammage. This has resulted in conserving the raw material. The loss was no more than 10 per cent.

Further, the use of hydrogen peroxide eliminates the chlorinated organic compounds and reduces the biological oxygen demand (BOD), as well as the level contents in the effluent discharged from the CMP plant. 90 per cent of the raw material is utilized and generation of solid waste (hyposludge) which is reduced from 6 mt/day (on dry basis) to about 1 mt/day.

Change over to hydrogen peroxide bleaching has substantially reduced the handling and usage of hazardous chlorine compounds. Chlorine requirement for newsprint at HNL was a mere 18.1 kg/mt in 1991-2 in contrast to 87.6 kg/mt if conventional sodium hypochlorite method had been used (C.M. Alias, personal communication, incharge, CMP plant).

A comparative account of the various discharges from the chemi-mechanical pulp mill with hydrogen peroxide and calcium hypochlorite bleaching is given below:

Parameters	Units	Calcium hypochlorite bleaching	Hydrogen peroxide bleaching	Reduction
1. Effluent flow	m^3/hr	441	389	11.8
2. Suspended Solids	mg/lit	961	610	36.5
3. Colour	Pt-Co	6,170	5,920	4.1
4. COD	mg/lit	3,050	2,480	18.7
5. BOD	mg/lit	1,170	780	33.3

Source: HNL.

It is not surprising that in spite of having these benefits, hydrogen peroxide is not used by the private sector paper industry in India. In the first place, hydrogen peroxide is very expensive. The problem for HNL is that it is produced only by one single monopolist—Bombay Dyeing. Its technology was completely imported and for HNL to get a licence for its own production of hydrogen peroxide from the government would be difficult.

Is HNL at the mercy of Bombay Dyeing for its supply of hydrogen peroxide? The plant in-charge claimed that the mill had its own calcium hypochlorite plant which was still retained as a standby and had not been scrapped. In case there was any problem with the supply of hydrogen peroxide the mill would have to switch over to calcium hypochlorite. This has naturally helped HNL to keep Bombay Dyeing in check (Field Work—HNL Plant in-charge—personal communication, March 1993). It is only now that Bhadrachalam Paper Boards Ltd., and J.K. Paper, from the private sector, plan to use hydrogen peroxide as a desperate measure of modernization in the face of the threat of liberalization.

Lease Agreement for Raw Material

HNL is at a disadvantage in relation to its counterparts in USA, Canada and Europe who continuously use a specific species of softwood. HNL has no choice but to use a combination of eucalyptus and bamboo in different proportions at different times. HNL started with a production capacity of 80,000 mt of newsprint per annum for which it was to be annually provided with 1,50,000 mt eucalyptus and 1,89,000 mt reeds/ bamboo. But, as stated earlier, the proportion changes.

To procure its raw material requirement HPC entered into a long-term agreement with the Government of Kerala on 7 October 1974. According to this agreement the Government of Kerala has to ensure a sustained supply of the following raw materials for a period of 30 years, according to clause no. 4(i).

(1) 1,50,000 mt of eucalyptus wood (at 50 per cent moisture level) every year consisting of 10,00,000 mt of *Eucalyptus grandis* and the balance in *Eucalyptus tereticornis.*

(2) 1,06,000 mt of reeds (with 10 per cent moisture) or 1,89,000 mt (with 50 per cent moisture).

It was also agreed that in case of destruction of reed forests due to gregarious flowering in the areas allotted to HNL, suitable long-fibred pulpwood from other areas would be supplied, as far as possible, in the agreed quantity (*Agreement*—HNL, dated 7 October 1974).

Presently the mill uses 50 per cent of *Eucalyptus grandis* and 50 per cent of *E. hybrid (tereticornis)* for chemi-mechanical pulping. One of the mill's experts on raw material claimed that *grandis* which has a larger girth than the hybrid was very little different from the latter in its fibre length, varying only in its basic density. Although HNL technicians are unaware of FAO's findings they found that this use of Eucalyptus hybrid is wasteful. It requires a higher amount of chemicals for separation of fibres and generates more coloured liquor and therefore requires more hydrogen peroxide for bleaching. The chemical requirement for bleaching of *E. hybrid* and *E. grandis* was 30 kg and 15 kg respectively. The mill's chemical engineer was of the opinion that an 8-year old crop is preferable to a 12-year old crop, since the older dry wood required more chemicals (C.M. Alias, CMP Plant in-charge—HNL, personal communication).

In 1984 HNL had sustained a loss of Rs. 8 crore attributed directly to non-availability of raw materials. It had to import 400 tonnes of chemical pulp at the cost of Rs. 5.5 crore and purchased raw material from private sources at an additional cost of Rs. 3 crore (*Economic Times*, 21 March 1985).

As the raw materials—eucalyptus and reeds—were being depleted fast, the mill used them judiciously, meeting a part of the pulp requirement with imported pulp. The different types of raw material provided by the Forest Department is given in Table 7.8.

TABLE 7.8: STATEMENT SHOWING QUANTITY OF FOREST RAW MATERIAL ALLOTTED TO HNL

(*in mt*)

Year	Reeds	E. grandis	E. hybrid	Bamboo
1982-3	38,896.170	85,453.850	14,434.340	—
1983-4	45,271.810	– nil –	– nil –	—
1984-5	38,852.470	53,149.770	– nil –	—
1985-6	57,330.620	78,995.270	11,256.200	—
1986-7	77,340.350	84,856.310	19,259.330	—
1987-8	61,354.820	48,083.870	1,862.640	—
1988-9	57,170.970	70,622.570	52,772.440	—
1989-90	72,694.360	76,774.695	15,655.380	—
1990-1	62,619.435	53,085.830	39,951.000	—
1991-2	64,480.435	75,080.485	34,563.554	—
1992-3*	54,223. 192	38,963.179	855.655	5721.455

Source: HNL.

Note: *From CCF office, Trivandrum, Letter no. G38191/93 dated 15 June 1993.

HNL Feels Threatened

Why was HNL, a blue chip public sector unit which was the highest producer of newsprint (NP) and whose quality surpassed others in 1993-4, suddenly saddled with unsold stock of NP? Why does it feel threatened'? The answer to this has to be sought in the Government of India's NP policy, its incapacity to straighten out the situation, and perhaps too, for its unconcern if not bias against its own mill. The threat is shared by Mysore Paper Mills and NEPA. Tamil Nadu Newsprint, with other options, would perhaps wriggle out. All the four units are in the public or in the state sector. Their total stock of unsold NP is 30,000 tonnes.

It is important to remember that the Government of India took on NP production only because the private sector evinced little interest in it, and also because of the recurring outflow of foreign exchange. Despite an enormous increase in NP production, it continues to import NP much in excess of the real demand. It has also shackled HNL and other NP units with various controls.

Even the requirement of NP is worked out annually by the Registrar of Newsprint for India (RNI), who functions under the Ministry of Information and Broadcasting (I and B). Its demand forecast invariably outstrips the actual consumption needs of the newspaper industry (NI), as shown in Table 7.9.

NI implicates petty dealers who through 'fenami' demands to the RNI, have inflated the demand forecast. They have also been accused of being involved with importing cheap NP from Bangladesh and China, and second grade quality of NP from the Soviet Block through rupee payment much to the annoyance of the NI. Good quality NP is also

TABLE 7.9: DEMAND FORECAST AND ACTUAL CONSUMPTION OF NP

(in tonnes)

Year	Demand forecast	Actual required	Actual consumption	Actual production
1975	3,00,000 NCAER		1,79,000	
1980	5,00,000 NCAER		3.65,000	
1985	6,00,000 Planning Commission		4,65,000	
1993		5,60,000		4,78,000

Source: Y.A. Rao 1989: 56.

imported from Canada and Sweden. The western capitalist countries have been facing a series of recessions. At such times, imported NP price falls below the indigenous variety. During periods of economic recovery not only do the prices go up, but the producers are also unwilling to export NP to India as happened in 1987-8.

According to the NP allocation policy, the requirement of NP has to be met to the maximum possible extent from domestic production, and only the residual quantity imported. But 'benami' demand has inflated the residual quantity to an enormous amount. The price differences between the imported and the indigenous increase during long periods of recession in the west. The reasons for the high cost of production of NP in India have already been indicated. The newspapers are business concerns. They therefore chose imported NP from the west which scores over the Indian only in terms of its price, the quality being almost the same. *The Times of India* claims that it has taught all the NP units who have willingly agreed to upgrade their quality to run on high speed machines. The Indian technicians had no prior knowledge of the needs of the modem press as this country has had no history of NP production. The imported variety from the west also scores over the Indian on two minor points, packaging and delivery and, to some extent, on quality control. The fault is not the industry's but the lack of basic infrastructural facilities like proper-sized rakes, warehouses, etc. These are typical problems of an underdeveloped country and not that of an industrial unit. *The Times of India*, for instance, does not buy imported paper during the rains. In other words, it has a stake in the survival of good indigenous industrial units. But as Indian business organizations tend to be short-sighted, NI manoeuvres to import the less expensive NP.

It is also important to understand pricing and allotment of both imported and indigenous varieties of NP. Prices are determined by the pricing committee of the Information and Broadcasting Ministry. In order to have some control over the imported NP it had to be canalized through the STC right from 1980 (till it was decanalized in April 1992, under pressure from NI). Incidentally, STC's NP scandal is finally being investigated by the Central Bureau of Investigation (K. Nag 1996). In March 1980, a customs duty was imposed at 15 per cent *ad valorem* which was changed to a specific rate of Rs. 825 per tonne in March 1982. In order to give relief to small newspapers (i.e. circulation less than 15,000 copies per day), no customs duty was charged. Also, a concession of 50 per cent duty was charged to medium newspapers, i.e. those with circulation between 15,000 and 50,000 copies per day. In 1984, the customs duty was reduced to Rs. 550 per tonne. The changing rate of customs duty is not relevant. What is relevant is the total lack of information about the number of small newspapers, and whether the 'benami' ones are being utilized to misuse the privilege of the small newspapers to collect and illegally sell imported varieties in the black market.

Since NP falls within the purview of the Essential Commodities Act of 1955, the NP control order regulates acquisition, consumption and sale of both imported and indigenous NP. The RNI is a functionary of the Ministry of Information and Broadcasting to purchase the consumer's pre-determined quota of NP. This quota is announced every year. The RNI is a convenor of the NP Advisory Committee (NAC). An *ad hoc* sub-committee has been constituted to advise the Government on the requirement of production and import of NP. It is interesting to note that the NAC sub-committee includes representatives of the Ministry of Information and Broadcasting, and Industry, of the STC, RNI as well as representatives of the newspaper industry. It is said that representatives of NI go abroad to meet the manufacturers of NP, bargain and beat down the price to the barest minimum whenever the Western NP industry is facing a recession. This explains why the cost of imported NP which was $807 in 1989 dipped to almost half in 1994. This price is far below the cost of production of NP in India, the in-flow of Western NP being regarded as voluntary dumping.

Duty on imported NP was abolished in July 1991, and now the allocation of 2:1 of indigenous versus imported is under pressure to be lifted—'Under the New Liberalization Policy it should be', says A.K. Das, Industrial Advisor on Pulp and Paper, Government of India (GOI), 'and perhaps will be soon. The Indian NP industry should learn to be competitive—that's Government's policy' (personal communication, March 1994, Delhi). But Kelkar's Committee, which is also a govermnent appointed committee, made basic recommendations in 1987: The indigenously produced NP should be consumed fully and imports treated as subsidiary measures, and that imported NP should have a duty element of 40 per cent. Newspaper establishment should be permitted to import any quantity of additional NP on this tariff. The small and medium newspaper should be allowed a maximum of 50 mt tariff free imported NP. Only as a long-term policy did it recommend a movement away from administrative pricing. Conditions have changed for such a movement. But the government chose to only abolish the import duty despite the strong recommendation against it by the Estimates Committee in their *Eighteenth Report* (10th Lok Sabha) on I and B newsprint policy in August 1992. It is also interesting to note that unlike as on NP, heavy customs duty on imported paper has not been lifted. Paper is produced by the largest business houses and the government cannot alienate them totally, despite ushering in the liberalization policy. The NP industry has to pay an additional central sales tax of 4 per cent.

The total stock of unsold newsprint in January 1994 was 30,000 tonnes. The Indian Newsprint Manufacturers Association estimates that the total production of NP in India today is 4.78 lakh tonnes after production has been opened out to other units, besides the basic four newsprint mills. The total estimated need is only 5.6 lakh tonnes, so it pleads in its letter to the Prime Minister dated 9 January 1994, that the obsolete ratio of 2:1 of indigenous to imported be revised to 5:1.

If the government does not take a decision, this spells a death knell for its own industry. Newspapers which are owned by big business houses realize the danger, and at certain points of time being wholly dependent on Indian newsprint, cannot rely totally on imported NP. The price of this can shoot up when the Western economy recovers as business cycles in the so-called free economy are uncertain and unpredictable. Even the petty agents who were initially responsible for wasteful import of NP are now powerful indenting agents of the NP companies who 'export' their commodities in India. One of them who is also an agent for one of the Indian NP companies also admitted to the grave danger of relying wholly on imported NP. Why then is the Indian government alone so insensitive to this issue?

TABLE 7.10: COMPARATIVE PRICE OF IMPORTED NP 1989-93

Year	Price of Canadian NP US $/mt)	Price of USSR NP (US $/mt)	Exchange rate of Re. per US $
1989	807	800	14.50
1990	620	610	18.00
1991	613	600	20.00
1992	571	550	27.25
1993	440	400	31.50

Recycling of Waste Water

PPI needs an enormous quantity of water. Much of it is wasted. The fresh water requirement of HNL is 44,300 cu.m per day for production of 250 tpd which works out to 180 cu.m per tonne of newsprint, drawn from the Moovattupuzha river.

The total quantity of waste water generated from the mill after use and recycling is about 37,500 cu.m per day, which works out to 150 cu.m per tonne of newsprint. Yet, HNL consumes much less water than any other PPI unit. The waste water carries colour although it is not a pollutant. From an aesthetic point of view it needs to be treated. HNL carries out an additional treatment for colour removal. This involves the use of alum, rare earth chlorides and polyelectrolytes (*Journal of Industry and Commerce*, 1993: 15). Adjustment of pH is necessary before primary clarification. The colouring matter is mixed with hyposludge. It is de-watered in a vacuum filter and disposed of as solid waste. The treated effluent water from the lagoons and clear uncontaminated sewer water from the mill is disposed of through gravity pipelines into the Moovattupuzha river through a dispersion unit extending about 45 m into the river.

Earlier, the system of effluent water disposal was greatly opposed by the local people. With the help of an anti-pollution committee in the year 1982, it moved the Kerala High Court for a writ directing the company to change its plans of discharge of effluents. The company feared that there would be a long battle ahead.

Farm workers, washermen, fishermen and housewives were among those who demonstrated outside the factory, demanding that the effluents be piped into the sea 35 km away, and not discharged in the Moovattupuzha river.

An action committee spokesman said a memorandum was submitted to the Prime Minister in February pointing out that discharge of the effluents into the river would pose health hazards and threaten the livelihood of about 1,00,000 people living on its banks.

A factory spokesman described the people's fears as baseless. 'A plant to treat effluents and prevent pollution has been established at a cost of Rupees four crores,' he said. The plant meets the specifications of the statutory central and state water pollution control boards.

He pointed out that the Moovattupuzha river had an estimated flow of 500 cusecs of water. The newsprint unit required only 20 cusecs and the effluents would be about 15 cusecs, which was 'insignificant' considering the flow of the river, he said.

Top management personnel were not willing to take in a mouthful of effluent water to show that it posed no threat to human, animal or plant life. 'Soon we will demonstrate to the suspicious people that fish can also be reared in this water', said one of them. He thought it would be a waste to spend Rs. 10 crore to carry the water to the sea ('Vellor Newsprint Project faces fresh snags', *Financial Express*, 19 June 1982).

Solid Waste Disposal

The solid wastes generated are coal ash (145 tdp), lime sludge (70 tdp) and clarifier sludge (10 tdp). These are disposed of in the low-lying areas of about 45 ha which are earmarked for the purpose. According to one of the employees, the above solid waste is first packed in polythene bags and then dumped in order to prevent its adverse effect on soil fertility. An area of 175 ha is provided for dumping solid effluents. Also the waste is disposed of systematically by dividing the area into blocks and the enclosures made for coal ash. The filled blocks are levelled by bulldozers and covered with earth to facilitate growth of vegetation. Plantations of different species of trees have been created all around the area.

Air Pollution Control

Air pollution is reduced considerably by installation of dust collectors for the emissions from the Power House, the Coal Handling System and the Chipper House. Provision of water sprays at these points to suppress the rising dust.

Electrostatic precipitators are used to arrest emission of gases from the use of large amounts of chemicals such as sodium sulphate, sodium carbonate that escape from the recover boiler.

Well-planned afforestation all-around the mill and waste disposal area have helped in the control of air and noise pollution.

Labour

Kerala is densely populated. It was expected that with the setting up of the HNL plant, employment would be created for the local people, leading to an overall development of at least Vellor in Kottayam district.

The mill was set up by clearing 695 acres and evicting the people living on this land. A question was raised in Parliament in 1986 (unstarred Q. no. 3315 dated 25 December 1986), whether the evicted persons had been compensated by providing employment. It was found that 272 families had been evicted while 232 persons from such families were employed by HNL. It was also stated that the maximum possible concessions and preferences were given to the evicted by the selection committee, which would continue to accommodate this category, provided the candidates were suitably qualified for the vacancies.

As in the case of all new PPI units, in HNL too, compensatory employment for the evicted is based on the number of families rather than on the actual employable number. The company has employed only 232 persons out of 272 evicted famllles. These statistics give no clue to the actual number of adults.

Secondly, the corporation promises to employ them depending on their qualifications. It is surprising to even expect these people to have any kind of technical qualifications. So the general trend would be to employ these workers for low-grade manual jobs for which they are found to be 'suitable'. The General Manager (Works) informed us that they were given indirect employment as contract and casual labourers.

Viewed from a wider perspective, HNL has failed to generate employment to the extent expected, mainly because it has employed a meagre number of evictees, and also because the scope for increasing employment for local labour is limited by its use of highly automated machines. Even if there arises any need for labour it will be only in the sphere of highly skilled and technical jobs. Also, the number of posts would be few, apart from the stiff competition. This leaves Vellor in its pre-HNL era. But HNL employs ex-servicemen and the handicapped, and in that sense, as the *Economic Times* (13 May 1983) puts it, HNL notches several firsts.

Facilities Provided to Labour

The mill complex has a large township providing modern housing facilities to about 1,200 employees. The township provides infrastructural facilities such as banking, postal and shopping facilities.

Informal discussions with mill employees provided a further insight to the available facilities. Only 50 per cent of employees live in the mill's quarters, mainly officers and managers. There is a central school but this is meant only for mill employee's children: the local population has no access. NEPA scores over HNL on this count. The reason for this was alleged to be a lack of demand for such a school. There is a primary health centre started by the state government, not by the mill. An HNL employee stated that there were a few schools and hospitals for the local population provided by the mill (March 1993).

Good newsprint is manufactured only in the public sector and in the joint sector. HNL is, on all counts, the very best of the four units. These units were set up in the core sector for India's rounded development, and also to reduce the outflow of foreign exchange. But the Government of India has not taken the latter seriously. Neither the reasons for the accumulation of the stock of NP, nor the danger it spells to an Indian core industry, are brought out in the open for public discussion. The major difference between the Indian and imported NP from Scandinavia and Canada is the price differential at this point of time (1994), when the West is facing recession, in fact a series of recessions; its price will shoot up when the trend of the business cycle changes. It behoves the Indian government to protect its fledgling industry and make it grow strong and healthy by reframing its policy.

SECTION 4

An Issue

The Changing Use of, and Conflicting Claims on, Bamboo Forests

It is only in recent years that some of the foresters concerned from the Karnataka and Kerala cadres have openly proclaimed that the traditional bamboo workers have a priority over the modem sector to the fast-dwindling bamboo stock. Such a position indicates more than an interesting shift from the earlier one held by the State Forest Departments in the paper belt; earlier they had abrogated their fundamental duty to guard the health of the natural forests. The question of priority needs to be examined in some detail; and if looked at historically would perhaps yield a workable answer.

To begin with, both the states, Kerala and Karnataka, have had powerful movements backed by social and natural scientists calling for a halt to the destruction of the natural bamboo forests by the powerful PPI. A few of the foresters, deeply troubled, have given their full moral support to the movement. The focus of the study at this point is Kerala and not Karnataka. Karnataka in fact has moved a step further by setting up a blanket ban on the felling of bamboos and other timber trees by wood-based industries. In comparison, the industrially backward states like Madhya Pradesh and Orissa have nationalized their forests, but continue to supply bamboo and hardwood to the industry through their Forest Development Corporations. Their priority is still industry over the traditional bamboo workers.

Before setting forth arguments in favour of the traditional industry, it is necessary to take a look at the bamboo scenario over a century and a half and the changing interest in it as an industrial raw material. The methodology of this study calls for this. For distribution of bamboo forest, see Part 2.

The natural history of reeds seems to give an impression that reed forests just a century ago were indestructible, and in fact proliferated with human interference in the forest ecosystem. Yet in recent years it is obvious that reed forests are under stress and are dying out, this causal factor pointing definitely to direct attack on the clumps through careless harvesting, flouting every silvi-cultural recommendation. Incidentally, the extent of reed forests estimated by the Govermnent of India Pre-investment Survey (1968) was 10,000 sq. km. In 1977 a reassessment showed that it had gone down to 9,400 sq. km. Later estimates are not available.

Use of Bamboo in the Traditional Sector

Bamboo had a special place in the compounds of large agricultural households. As aggressive claimers of space and nutrients, bamboo's place was always near the hedge (K.K. Seethalakshmi, personal communication, 1993). People need a mature culm or two, perhaps even more culms would be cut down to mend rafters and fences, to provide props for banana and other weak-stemmed plants. Such households also needed baskets and mats, some of them for storage. Strips of reeds were more malleable than bamboo for weaving mats; besides, reeds were available in the nearby forests, and a necessary number of mature reeds alone had to be carefully cut from these forests taking care to see that one-year and two-year culms remained undamaged for the next year. It is important to remember that weaving was undertaken by the landless agricultural labour attached to a wealthy agricultural household. They were either scheduled castes or tribals. Such work was an off-season activity and undertaken as a part of feudal obligation to the landed aristocracy (Nair 1986: 100). With the expansion of trade, the reed industry grew, enough to cater not only to the local markets but also for bulk needs of urban centres. Due to its low cost and durability, mats began to be used as dunnage material in the warehouses even as far back as a century ago and were exported to trading centres (Bourdillon, 1893, quoted in Nair, p. 100). Large-scale production of baskets was started to pack fruit, vegetables, betel leaves and fish. The wartime demand for mats, for putting up temporary shelters on the front, rose (Ajit Kumar 1985, quoted in Basha).

Numerical Increase of Bamboo Artisans

Profitable use of mats caught the attention of the intermediaries. They entered the traditional sector and made good profits by purchase and sale of mats. Realizing the importance of economies of scale, bulk production was started in and around Kalady and Angamaly in Ernakulam district. The choice of the place was not fortuitous. Availability of reed in the Periyar river basin and feasibility of water transport were important factors (Nair, p. 100). Though the demand dwindled after the war, it again rose in 1958, when a large number of sugar factories in India used bamboo mats for refining brown sugar. This encouraged a number of people including those from the forward communities to learn the art of weaving from the *sambaras* and *purayas*, the original weavers. The newcomers have made weaving their vocation, to earn their livelihood. The number of weavers therefore began to swell.

Promotion of Traditional Sector

The poor unorganized sector of mat-weavers reeled under the grip of the ruthless merchants who monopolized both extraction and supply of reeds to the weavers and also bought their finished products. In order to enhance the income of the weaker section, viz., the mat-weaving people, the Kerala government first constituted a cooperative society to channelize marketing of their products. But the action was inadequate to protect the impoverished bamboo workers. Finally, in 1969, a Bamboo Enquiry Committee was constituted and on the basis of its recommendation, the Kerala State Bamboo Corporation was set up on 13 March 1971. As a first step, the Corporation took up only the purchase of mats, opening out new outlets for them. As a second stage, the corporation entered into reed extraction to free the traditional weaver from the merchants. Both extraction and the supply of reeds became the monopoly of the corporation. Initially, the corporation bought the reed from the Forest Department; later the government waived payment in order to promote the traditional sector, and support the poorer section. But the unregistered workers continue to collect their own reeds and pay the government for them without receiving any benefits; this is because the corporation cannot reach their isolated homes, while such homes are spread throughout the state.

In bulk production areas, Angamaly and Kalady, the weavers get nothing more than wages. In still another sector, the cooperative reeds are supplied to the registered members on realization of actual collection and transport charges. In other areas too, registered weavers are given reeds on the same basis as those in the cooperative sector (Basha 1992: 15-18). By and large, the benefit to the traditional sector is in terms of regular wages. C.T.S. Nair (1983) estimates the annual income of the mat weaver to be a paltry Rs. 820 per annum. Why then do the foresters call for the perpetuation of this industry?

The number of traditional bamboo workers in rural Kerala has now swelled to a enormous 30,000 (Nair 1986). *The Times of India* (28 June 1993) puts it down to 71,000 which includes both natural growth as well as penetration of outsiders into it in search of employment. He reports that basket-making and mat-weavmg provide the maximum employment within the forest industry sector. The reasons for their penury, despite government help and despite the utility of their artistic wares, must be explored fully and proper solutions suggested.

The answers have to be sought in the polity, not just of Kerala in isolation, but in the larger context of the national setting in terms of the central government's planning for development. In this context its relationship with Kerala and its indirect control over the forest sector must be considered.

A Brief History of Kerala State

Immediately after Kerala was constituted as part of linguistic state reorganization, popular communist government was elected, to the discomfort of the Congress which was in power at the centre. Kerala's wealth lay in its vast forests and its labour power. The state set to work on land ceilings and brought in a relatively more equitable distribution of land, but the vast plantations remained untouched. The productive agricultural land is a narrow strip totally insufficient to provide decent living standards for the population. Encroachments into forest land were permitted and some even regularized to ease the situation, on an average the family holding land in the encroached areas never exceeding 1 acre. But industrial development and even building of the infrastructure for Kerala's growth was thwarted. The big business in India, in close collaboration with the centre, could not tolerate the development of a Communist or even a Left Front state. But those business houses which manufactured pulp and paper needed bamboo—the long fibred cellulosic material—for their production and profit.

Entry of the Modern Sector—The PPI

The popular and even the academic notion that the Kerala government invited the three PPI units into the state in the hope that they would generate employment, if not development, has not been borne out by facts. Their entry into Kerala needs to be looked at carefully, even if briefly. Dalmia was one of the first five to penetrate into British monopoly control over the paper industry in the 1930s. In Kerala, he bought over an ancient British reed-based mill, the Punalur Paper Mills, in 1968 which was originally built in 1890 at Punalur. The mill's annual consumption of reed used to be 750 tonnes which it extracted by clear felling, totally disregarding the need for either rotation or selection felling. As a result the reed forests of Thiruvananthapuram, Punalur and Thenmala were damaged beyond repair. The area for reed extraction was then shifted to Adimalai and Pooyamkutty as the mills' requirement also rose from 33,000 tpa in 1972 to 50,000 in 1975 and still further to 85,000 tpa in 1982 (M. Chudamannil 1990), it finally closed down in 1986. For Kerala, the working unit meant a loss of flora, the valuable reed which provided protection to the fragile evergreen forest ecosystem; and a closed unit implies a further burden of unemployed labour force. To the industry saddled with its obsolete machinery production, this meant a loss of profits through its inability to compete; closure on the other hand meant a gain.

The corporation was able to repeatedly raise loans at subsidized rates from the public sector financial institutions under the special clause for modernization of a sick unit. The huge losses are for the financial institutions which are compelled to write off the loans as bad debts; there was no loss for the corporation.

Birla, Dalmia's contemporary and an early rival in paper production, had forged forward and diversified from bamboo-based paper to rayon, a fabric which was much in vogue in the 1950s. Setting up of a large 72,000 tpa of Gwalior Rayon called for a larger catchment area. The rich Nilambur forests in Kerala seemed an inexhaustible source, and Birla did get it, but evidently not directly from the state government or its Forest Department. The central government had probably to intercede, perhaps impose. There are no ready records to show what transpired between the central and the state governments. But the first, as well as the subsequent agreements for the contracted areas for bamboo, give an impression that Gwalior Rayon (GR) was imposed on Kerala by the central government rather than invited by the state government.

In 1959, the Communist government in Kerala was dismissed by the centre on the recommendations of the Congress President, Indira Gandhi, on a flimsy pretext: On 3 July 1958, the first contract for the assignment of the catchment area of raw material was assigned not between the representative of the state government, or the Forest Department and the Corporation, but between the Governor of Kerala and the Corporation. The Governor is not a representative of the state but of the centre. The terms of the contract and its tone are harsh and demanding. It was never so blatant in any other state. The agreement gave the company exclusive rights to fell bamboos from the well-stocked Nilambur forests. The contract, however, states that if there is a shortfall in the anticipated annual yield of 1,60,000 tonnes, the company has the right to move into contiguous forests; this provision was made on an anticipated rise of a 100 per cent increase in production. Birla's experience with virtual clear felling bamboo forests elsewhere, dates back to 1937. The royalty rate of bamboo is fixed at a token rate of Re. 1 per tonne. States vie for development, particularly industry which is seen to generate both employment and wealth. It is doubtful if any state government would be willing to sacrifice its precious resources and the revenue it would fetch, if sold at market price. The foresters throughout the country are bitter over the secretiveness and the political deals over the fixing of royalty rates at the expense of their forests (personal communications, also see K.P. Sagariya [1971: 557-8]; Waheed Khan [1971: 558-61]; P.S. Kothari

[1971]). Nevertheless, with the exception of Gwalior Rayon, now Grasim, it is the state government that sets the rules and not the corporation.

The first agreement also permitted a rent-free lease 'of site or sites for erection of store houses, sheds, depots, agencies and other buildings of a like nature, bona fide required [*sic*] for the purpose of business'.

In the second agreement which was entered into within four years, on 6 August 1962, the contracted area was stretched even further into the reserve forests in Wynaad, Palghat and Nenmara. The second supplementary agreement signed on 10 July 1974, admitted depletion of bamboo and that the yield had fallen to 40,000 tonnes per annum, the agreement stated that the shortfall therefore was to be made up from outside the contracted area at a higher royalty rate which would range from Rs. 18.36 per tonne depending on the time-lag between felling and weighment. In the old contracted area, the rate remained almost unchanged: Rs. 1.67 per tonne if weighment was made on or before the 30th day of felling but before the 75th day, the highest, Rs. 3.75, if after 75 days. According to the agreement, shortfalls were also to be made up by eucalyptus at Rs. 25 per tonne. This again is a heavily subsidized price. The third modified agreement on 20 November 1976 raised the quota of bamboo and eucalyptus to 2,00,000 tonnes. It is referred to here only because of some unusual clauses that are detrimental to the bamboo forests and which no state Forest Department would tolerate. The agreement states that fresh areas would be allotted on 1 June, and the company must complete its felling by 1 July. This is an amazing concession, a gross transgression of all silvicultural felling rules of bamboo which forbid anthropogenic interference from June to September, when new bamboo culms grow at a rapid rate and reach their full height, but not their strength. June to September is recognized as a vulnerable period when bamboo forests all over India are closed. The agreement also makes it clear that the bamboo pulp should not be used for anything other than what was contracted—rayon, but the balance sheets of the company indicate some production of paper in later years.

The agreement dated 27 November 1988 states that the company has ceased to function since July 1985, and also the insufficient quota of eucalyptus had to be made up with a supply of *Bombax malabarian*. Dalmia's paper mill was somewhat sore at the differential royalty rate granted to Birla's rayon and to its own Punalur Paper Mills (PPM). The latter was charged at Rs. 18 per tonne of reed, when at the same time, in the early 1970s Birla still paid only Rs. 1.67 per tonne of bamboo (see PPM's memorandum presented to the Kerala government on 24 January 1976), and at the differential standards used for measuring a notional tonne of eucalyptus (see PPM's letter dated 15 January 1977), addressed to the Additional Secretary, Department of Agriculture and Forests, Government of Kerala. And the last straw that upset Dalmia was that a 200 tonne newsprint mill as well as a pulpwood plantation were established for this purpose. This was in response to the central government's team which visited Kerala in 1971 (Letter no. 2-11/71 dated 16 August 1971, from Ministry of Agriculture—GOI, New Delhi). Both Punalur Paper Mills and Gwalior Rayon, the latter in particular, have fleeced the state exchequer, denied its meagre revenue, ravaged the bamboo forests and disturbed the fragile forest ecosystem. They should be compelled to pay full compensation for the irreparable damage they have caused. The central government, which engineered the entry of the Birlas into the state, shares equal responsibility for impoverishing a state and should make amends for its loss.

Kerala has one more paper unit—The Hindustan Paper Corporation (HPC). It is not clear whether it was imposed on Kerala or welcomed by the state government. A 1971 document makes a stray mention to the effect that the Kerala State Industrial Development Corporation had proposed a joint venture to establish a newsprint industry. There is no indication as to what the response was. HPC is a public sector corporation set up by the central government as a standby in the 1970s, when the powerful private sector paper corporations got together to defy the government and gave up their contractual obligation to produce inexpensive writing and printing paper for the weaker sections. The private sector PPI unit had, by and large, quietly renovated their machinery, in corporate terminology 'modernized', to produce highly profitable speciality and industrial paper. Elsewhere, it has been explained that the central government had as an additional standby encouraged small entrepreneurs to produce paper for common usage from bagasse and other agricultural residues rather than directly enforce and penalize the PPI. The threat of Emergency, rather than the government's Price and Production Control Acts, compelled the PPI to produce their quota of socially useful paper for a brief span of time in 1975-8. This produced a glut after which the 'cartel' began to defy the government. But in the early 1970s there was a shortage of newsprint. The agreement to provide infrastructural facilities, electricity and water, and to make available the requisite quantity of bamboo and eucalyptus as raw material, to set up HPC in Kerala, was again signed by the Governor (hereinafter called the government) and the representative of the HPC on 7 October1974.

Apart from the signatories there are interesting differences between the two agreements. To begin with, the HPC agreement states that 'the Government of Kerala in pursuance of the policy of Industrial Development of the State are [*sic*] willing to promote and encourage the establishment of the project in the state and in consideration agreed to grant and vest in the Corporation concessions, rights, privileges and benefits herein mentioned'. The concessions given to HPC do not in any sense measure up to the concessions bestowed on Birla, or rather those grabbed by Gwalior Rayon (Grasim). The raw material for HPC was 1,50,000 tonnes of eucalyptus at a royalty rate of Rs. 11 per tonne and only 1,05,000 tonnes of *Ochlandra*, the prevalent type of bamboo in the region, at Rs. 12 a tonne. At this point the concern of this study is only with the differential royalty rates for bamboo for the units. An unfair deal for the public sector, a PPI or otherwise, is quite familiar (see report on NEPA). HPC's unit in Kerala was mainly meant to produce newsprint and will therefore be referred to as HNL. It has been singled out for a special type of pricing of raw material, it is linked with the pricing policy of newsprint it produces, this is apart from the fact that the basic cost of bamboo is 12 times higher than what was set for Grasim and had remained so in 1974. HNL's agreement alone carries a penalty clause: an additional Rs. 25 per tonne besides Rs. 5 per tonne as development tax and an additional sales tax has to be paid by the corporation, if fire occurs in a coupe during the working period. Penalty is also to be levied if the 'society's forest wealth in a working coupe is lost or degraded'. Even if these clauses arise from the wisdom of hindsight, they have not been introduced in the second supplementary agreement of 1974 with Birla's Grasim. On the contrary, the contract permits extension of its leased area on the ground that bamboo in Grasim's catchment area had been degraded and the annual yield has fallen from 1.6 lakh tonnes per annum to a mere 40,000 tpa. Grasim's destruction coupled with its growing demand for fresh bamboo forests kept apace, but the state government was apparently helpless.

The Kerala government evidently did not willingly approve of the low royalty rate. In order to safeguard its forest wealth as far as possible, it passed Act no. 29 of 1978—The Kerala Forest Produce (Fixation of Selling Price Act 1978 [Notification dated 26 September 1978]). This act was evidently meant for a proper regeneration and maintenance of forests in the state; which is not possible unless the produce is sold at reasonable rates, and the proceeds utilized for that purpose. It noted that timber as well as bamboo and eucalyptus were sold at unreasonably low prices. The Act was to apply to the entire state of Kerala and to come into force at once. The prices were to be fixed at the end of each financial year and determined by an expert committee which included the Chief Conservator of Forests as well as the Director of Kerala Forest Research Institute, besides the Secretaries to the Department of Industries and Finance. Ten per cent of the selling price of the produce was to be set aside for the development of forests.

The Case of HNL

Now to the crucial question, has HNL any right to exist? If yes, what is its position vis-a-vis the bamboo forests and the ecosystem? Technologically viewed, the HNL unit in Kerala is eco-friendly; with the chemi-mechanical method of pulping it utilizes as much as 88-90 per cent of the cellulosic raw material, 95 per cent perhaps if it had not been saddled with FAO's eucalyptus. Contrawise, the loss of raw material is only around 10-12 per cent. This contrasts very positively with each of the private sector PPI units which persist in the outmoded and wasteful sulphate process, which yields no more than 38-40 per cent of the cellulosic material while the rest of the 60 per cent or more is discarded as waste. Modernization implies very heavy investment. The private sector is interested in profits, not bamboo forests. HNL is perhaps the only unit in India which uses the expensive but safe hydrogen peroxide for bleaching, unlike the private sector units which use chlorine, a dangerous and environmentally polluting agent. The HP Corporation feels threatened, as the Wadias hold a monopoly control over the manufacture of hydrogen peroxide and can freeze HNL if any dispute should arise.

Setting up of this extremely modern, highly automated, public sector unit was the central government's panic reaction to a crisis situation, a second paper famine created by the PPI cartel. Punalur Paper Mills was part of the cartel, yet it had the audacity to propose to the central government that it would undertake production of newsprint [see letter to the Chairman of the HPC (19 May 1977) and to the Ministry of Industry (26 May 1976)]. HNL had to be modern to ensure an uninterrupted flow of writing and printing paper as well as newsprint, especially since the central government had neglected and ruined NEPA. Introduction of new pollution-free technology was a result of continuous pressure of environmentalists the world over. HNL benefited indirectly from various distant pressures. Kerala was still rich in bamboo in the early 1970s, but since it is also known to have a labour force highly conscious of its rights, automation perhaps seemed a way out. Naturally, therefore, the employment generation potential of HNL unit is limited. The factory records put down its labour power to 1,800 in 1993 in contrast to 3 lakhs in the entire traditional sector (Government of Kerala 1983, quoted in C.T.S. Nair 1986). Another 5,000 are estimated to be employed for collection of reed for the traditional sector. Another 2,000 to 3,000 are employed in transport of reed, supervision and management. Improved accessibility and growing unemployment have drawn even women into this activity, particularly basket-making and mat-weaving (Nair 1986: 102).

Despite the poor employment generation capacity of HNL we hold that HNL which was initially set up to produce socially necessary printing and writing paper, and now newsprint, must be supplied with adequate raw material, since newsprint is necessary for rounded development.

HNL uses both bamboo and eucalyptus as its raw material. The team also believes that bamboos with their dual systems of propagation by natural reproduction and regeneration and multiple systems of vegetative propagation through human intervention are ecologically the most favourable species. They are a naturally renewable resource provided they are treated with respect and harvested with care. Bamboos can be propagated if necessary on a very large scale and introduced in place of ecologically harmful and even doubtful species like the eucalyptus. Much research on vegetative propagation, conventional as well as modern tissue culture besides the techniques of seed preservation on species specific to the Kerala's ecosystem, have been carried out by K.K. Seethalakshmi, T. Surendran, C.S. Venkatesh and T. Soman at the Kerala Forest Research Institute (KFRI) at Peechi. In fact, the KFRI publishes a bi-annual *Bamboo Information Centre Bulletin* which provides excellent abstracts of research on bamboo in South and South-East Asia. It would be pedantic to list the numerous successful projects on propagation of bamboo and on bamboos as plantation species.

Bamboo and Reed Felling Practices by the Traditional and the Industrial Sectors

In terms of manufacturing socially useful products and in its use of environmentally safe technology, HNL has not only a right to exist but to grow and flourish. However, in terms of preservation of its natural raw material—the reed, its rating is low. It has to be rated vis-a-vis the felling practices of Kerala State Bamboo Corporation. Setting up of this corporation to provide free raw material for the traditional sector by the state government is commendable, but are its felling practices superior to those of HNL? HNL's felling practices have also to be rated along with the individual mat-weaving households scattered all over the state, but who fall outside the corporation's network. What were the reed and bamboo collection practices of people earlier?

Two centuries ago the pressure on the bamboo/reed forests was nominal. Both the tribals as well as non-tribals who were dependent on forest raw material for constructing their huts or for making baskets or weaving mats for their own use or as feudal obligations would cut only a few mature culms for their own immediate use, from the vast forests close to their own homes. The pressure did not increase to any dangerous level even when the bamboo products became commodities for the local market. But it increased steeply during and after the two World Wars when the bamboo products reached the national markets, and more recently after they were used for making decorative boards by the Bamboo Corporation. Due to the high tensile strength of bamboo, much successful research is being carried out on how to also use bamboo as structural material (see all the issues of *BICB*). The potential threat to the bamboo forests from the organized traditional sector is therefore growing rapidly.

Since 1977, the Bamboo Corporation has almost a monopoly felling right over bamboo/reed to supply the traditional sector. It engages a large number of knowledgeable as well as novices to cut the reed. They swarm the reed forests and resort to faulty techniques for two reasons—first, in order to maximize their earnings, the workers selectively fell every one of the mature culms and leave none as support for the tender new ones. Secondly, in order to reduce their workload they fell clumps closest to the loading points, thereby avoiding carrying or dragging the felled material in a difficult terrain from the interior areas. This sort of concentrated felling has rapidly depleted the approachable areas. To make matters worse, reed forests allotted for the traditional sector are even worked during the closure period (Basha 1991). True, mat-weavers and basket-makers require raw material round the year but the closure period—June to September—is also the growing period for bamboo, whatever the variety. They grow to their full height in the brief three-month period, but remain thin and brittle and break at the slightest touch. Damage is inevitable when the felled culms are dragged out of the tangled clump. This eventually amounts to clear-felling, which takes its toll of the rhizome and eventually results in the death of the whole clump. This need not have been so if simple rules of felling had been resorted to and the monsoon season observed as the rest period. Adequate quantity could have been felled earlier. Soaking of felled bamboos and reed culms in the streams, ponds, backwaters should presumably be no problem in Kerala, but we have no studies to substantiate this (Sulthonil 1990). Simple and cheap preservatives are also available (Dobriyal 1988). The Government of India's office of the Development Commissioner, Handicrafts, Agartala, has brought out Technical Data Sheets on *Low Cost Treatment of Bamboo for Handicrafts* in 1986. These deal with the prevention of both insect and fungal attacks.

The felling practices of HNL are even more damaging mainly because large areas are allotted to it, and also because its felling is restricted to areas close to the road. A large number of reed-cutters with their families throng and clear-fell entire clumps. Unlike the artisan, the paper industry consumes reeds of any age. Since payment of wages is based on weighment and the young reed is water-bound and heavier, the cutter sees no reason for sparing them, especially when they increase his earnings (Basha 1991). The difference between absolute clear-felling for HNL and the non-scientific selective felling of every mature reed by the workers of the Bamboo Corporation (BC) is only academic, for the tender yearlings cannot withstand the rough handling by BC labour and entire clumps are damaged. The only real difference is that the quantum of reed required by the traditional sector is appreciably lower than that needed by the PPI. Consequently depletion and damage to the reed by the most modern HNL is quantitatively higher. Viewed from the traditional sector which needs bamboo all the year round, the Forest Department accommodates it through an overlapping working system. Both agencies fell or work the same area at different times of the same year (Basha: 19-21), the result is disaster.

The Status of the Traditional Bamboo/Reed Worker

Perhaps the most sophisticated argument in favour of the traditional sector and against the modern sector is put forth by C.T.S. Nair (1986). Besides arguing in terms of differential employment generation and felling practices of the two sectors, he points out that the modern sector unlike the traditional, consumes enormous quantities of power and water, and also argues in terms of their differential investment of capital and their skills. The investment of an organized household worker never exceeds Rs. 15. All that an artisan needs is a good bill hook, a sharp knife or a blade. The Bamboo Corporation's investment in a traditional worker is a little higher—around Rs. 50—as the corporation maintains a number of collection depots and warehouses. The investment rate rises to Rs. 85 in the case of workshed production. The most sophisticated wares of the artisans are a variety of table mats which need simple weaving looms and storage space. Even at this extreme end, the fixed capital for a worker seldom rises beyond Rs. 200. Nair argues that the skill for weaving can be easily acquired by children but this spells danger of child labour; yet it is the women who form the major workforce. Nair however argues that they can combine weaving with household chores. It is difficult to believe how such a sensitive person can overlook the burden of dual duties on the women. The fact that women do it out of sheer need for survival is quite another matter. The fourth and the most important question is that of wages. Nair (1986: 102) himself shows that as the weaver earns on a piece rate basis, the artisan's wages dip to the level of opportunity cost, an average of Rs. 820 per annum. Reeds have also to be collected before they can be worked upon. Reed collection is a seasonal activity, although a reed collector gets a higher Rs. 40 per day income in terms of average being only Rs. 3,000 per annum. The Bamboo Corporation has a permanent managerial and supervisory staff whose average annual wages rise to Rs. 7,500, but they are not involved with production of baskets, mats, etc. There is also a private sector which controls the basket-makers and the mat-weavers; the managerial work is taken on by the entrepreneurs, Nair confesses it is difficult to distinguish between profits and wages; the nature of production relations in the last category is not spelt out by him.

The reed-cutters in the modern sector earn no more than those in the traditional, but the factory workers earn ten times more than the basket-makers and reed-weavers, and the managerial staff twice as much (C.T.S. Nair).

Structural Weakness of the Traditional Sector

Nair mentions a disquieting but not surprising finding. Both the household and the cooperatives, faced with shortage of capital, turn to traders and intermediaries and in doing so, get caught in their stranglehold (p.107). Nair also mentions that due to the dispersed nature of production and low value-added generation in the traditional sector, the multiplier effect is very low; in this sense it contrasts sharply with the modern sector. He explains the high multiplier effect of the modern sector in general in terms of their forward and backward linkages. The foreign exchange earnings of the basket/mat-weaving sector is very low, whereas the import substitution of paper and news-print is high (Nair 106).

Nair mentions a few other negative features of the sector he supports: cooperative units he finds tend to be inefficient and a large number of reed-processing units have closed down (Nair and Murlidharan 1983).

Plastic sheets and plastic household ware are a strong potential threat to reed mats and baskets. There is little doubt that bamboo is aesthetically pleasing and bio-degradable. On these two grounds alone, bamboo scores over cheap plastic wares. But what matters to the large mass of people is not aesthetics but price, and inexpensive plastic ware as a commodity scores over the beautiful bamboo ware.

Wishful thinking apart, the traditional bamboo industry is weak. An attempt to prop it up by setting up a Bamboo Development Corporation has not really helped the industry to stabilize and grow. If one goes by the average wages of the bamboo/reed artisans, as many as 30,000 by Nair's 1986 estimate are condemned to dismal living standards. Now *The Times of India's* (1993) estimates of bamboo workers is 70,000. Unless there are some very radical structural changes which will recognize basket-making and mat-weaving as honourable well-paid work of artisans, there is no redemption for them. No forester or scholar has a right to argue for perpetuating the system. Their arguments have to go beyond asking for the monopoly control of the reed/bamboo. Their position, no doubt a compassionate one, would neither help the reed workers nor the reed forests.

Need for a Fresh Historical Assessment

The scholar-foresters have viewed the problem statistically. If they had cared to assess it historically and within the national perspective, they might have located the factors that hampered Kerala's industrial development. Industries would have opened out varied avenues of employment for the naturally growing population, and then proper weightage would have been given to the question of traditional versus the modern sector's right to the forest raw material.

The most important factor to take note of is the composition of the ruling class at the centre, and from what follows the centre's relationship with the state. From the point of view of the study, we look at only one set of relationships, namely, the centre's growing control over the state's forests—its constitutional prerogative and the centre's linkage with the core sector industry which controls PPI.

The Government of India-Industry Nexus

The Congress was elected to power at the centre with the full support of Indian big business. But before doing so the latter set the ground rules for industrial development (see Bombay Plan 1945). Since our concern is with the production of paper, it is important to remember (as mentioned earlier) that a decade earlier, four of the most powerful business houses from the eastern sector had with great courage and and cunning broken the British monopoly hold over the manufacture of paper and entered the industry when the latter's power was weakening. World War II and the difficulty in importing essential industrial goods, of which paper was one of the items, gave a boost to the big business houses to profitably enter into production. And with state support and patronage the industries grew after the transfer of power. Production of printing and writing paper was given primacy—the Planning Commission saw it as a means to serve the social good. The late 1940s were the heady days of Nehruvian socialism, which announced the need for free and compulsory education for all. But there were other formidable impediments for rounded development and those had to be surmounted. In the countryside the *zamindari* system had to go in order to release the productive forces. And it did. But it was not sufficient and compromises had to be made. Feudal elements that controlled the vast countryside with its enormous population were not alienated. It would have been foolish to do so in the context of electoral politics. As a result the land reform policy that was proposed, over the years was watered down beyond recognition into tenancy legislation.

Even so, Kerala was the first to briskly enact the land ceiling act.

But it had so little agriculturally productive land, and the land–person ratio was so disproportionate, that it was impossible to make Kerala an agriculturally prosperous state. For the same reason it was impossible to provide a means of decent livelihood—or plots of land for the vast attached landless agricultural labour. As a survival strategy, some of the landless agricultural labour turned to their feudal off-season activity of basket-making and mat-weaving, making this their main source of income. As already stated, setting up of cooperatives and a Bamboo Corporation helped to act as props, but these have not helped the deprived bamboo workers to find their feet and they continue to survive below the poverty line. True, there is wealth today, the petro-dollars created by Kerala's labour-power toiling outside the country. In the 'free' economy of Kerala, this wealth was not put to productive development so that it has resulted in uneven prosperity. Before the forest question is entered into, it is important to remember that PPI's raw material is forest-based and that it emerged as an oligopoly with tremendous clout. Two members of the oligopoly were in Kerala. They were the Dalmia's Punalur Paper Mills and Birla's Grasim. HNL is a public sector unit set up by the central government as a very modern reliable standby whenever the private sector turned treacherous despite the government pandering to its needs (see case study of any of the private sector units). HNL was never meant to be a counter-weight—it is a standby.

The Forest Question and Centre–State Relations

Post-1947 India did not make its 1952 forest policy appreciably different from the colonial policy of 1894; some of the old pressures remained, and new ones were added in this changed situation. As in the earlier phase, again, after the transfer of power, Forest Departments in the elected governments in the states were expected to have two important responsibilities, namely (a) meeting the needs of the agricultural sector and (b) enhancing the government's income so as to meet the growing expenditure on development works. These have had a direct impact on both forests and forestry (FAO 1984, p. 53). Due to the central government's intervention in determining the royalty rate and demarcating the felling areas for PPI especially for Birla, Kerala had much to lose.

The 1952 National Forest Policy merely provides a broad framework applicable to the entire country. In theory, the most important documents for the foresters are the working plans, written schemes aiming at the continuity of policy and action at the divisional level, dealing with technical, operational and financial aspects of forest management. But over the years the situation has changed. Superimposed over the working plans rather than parallel to them are the priorities set by the Five-Year Plans. The plans are not mere guidelines on national priorities but set targets and also provide investment funds. These are applicable to the forest sector and are a great temptation, and perhaps the only source of funds, for forest activity. Royalties for raw materials have not enriched but drained the forests and its coffers dry. In other words, the working plans have receded into the background, though forests are constitutionally declared a state subject.

In the immediate post-War era when the PPI was surging forward, bamboos were plentiful. So the question of raising pulpwood did not arise. During the first and second plan period, there was a woeful shortage of fuelwood and matchwood in India, so Kerala dutifully took on these two programmes of starting fuel and matchwood plantations as the central government provided subsidies. The first and second plans had made allocations for them (GOI first and second Five-Year Plan and FAO 1984: 58). And it was not an Indian but an FAO expert, Von Monroy, who anticipated shortage of raw material for PPI in India and made suggestions which were to cause irreparable damage to the natural forests in India. His first two suggestions were made in the 1950s and were incorporated in the third plan. He recommended: (a) 0.6 Mha or 1 per cent of the most productive forest zones to be selected for a ten-year programme (1965-75) of plantation with fast-growing species at the rate of 60,000 ha every year for intensive management to ensure 4.5 million tonnes of uniform industrial material from 1975 onwards. (b) He also recommended changing the pattern of consumption of PPI from predominantly bamboo to predominantly hardwood with long-fibre component. Immediately, in 1960, with the central government's advice and resources, a grassland afforestation division on the *sholas* and an industrial plantation circle by clear felling of evergreen forests were set up by KFD (FAO 1984: 52). Eucalyptus plantations which accounted for less than 1 per cent of total plantation in 1960 in Kerala rapidly increased to 25 per cent by 1982 (FAO 1984: 52). The rapid increase again is because the National Commission of Agriculture, which the centre had commissioned, was asked to make specific recommendations to strengthen forest based industries with forestry. The fourth plan's objective was to achieve self-sufficiency especially for pulp and newsprint. NCA (1969: 109) in its special ninth volume echoes FAO's recommendations and to make it feasible for setting up a special wing called the Forest Development Corporation (FDC) in every State Forest Department, finally exhorting the premium financial institution, the Reserve Bank of India (RBI), or rather its agricultural wing, the Agriculture Re-finance Corporation (ARC), now the National Bank for Agriculture and Rural Development (NABARD), to make resources available to fulfil the new 'ideal'. The enthusiasm was infectious; Von Monroy revised his opinion and now recommended 2 per cent of the most productive forests to be converted to pulpwood plantation. Drought and famines in the early 1940s were ignored and the fifth plan, not to be outdone by any FAO expert, declared the aim of the 'dynamic' programme of production forestry to be achieved by clearing human-created forests with institution finances. Human-created forests were to cover 2 per cent by the end of the plan period. After all, NCA (1976: 4), had declared 'The requirement of industry should be the major reason for initiation of forestry activities.'

And so it was till the people began to spontaneously react. The most popular revolts are of course the Chipko in the sub-Himalayan region, Apiko in Karnataka, and the movement in Bastar to prevent the World Bank (WB) and the United States Agency for International Aid (USAID) programme to clear fell the natural forests of Bastar to make room for pulpable Caribbean pine, to name just a few prominent ones. The popular movements in India were the spontaneous protests of the local people to re-claim their natural rights to the forests that were progressively lost and their demand for at least a mimmal share in the forest produce. Social and natural scientists, sensitized by repeated publicity given by the Western Green Movement, joined their own people and provided the necessary leadership.

The upshot of the numerous struggles was the newly 'sensitized' Indira Gandhi and her gift, the powerful Environment Ministry which was set up in 1980. It was meant to keep surveillance over the profligate Forest Departments which enthusiastically or otherwise permitted PPI and other wood-based industries to ravage their forests. That the FDs followed the centre's dictates is another matter. It is important to remember that the pretext of the forests being a state subject ceased, and in 1977 these were put on the concurrent list. The same year, Social Forestry (SF) was tagged on to the Forest Departments, while Forest Development Corporations were ordered to be set up to clearfell the most potentially productive forests to make way for pulpwood plantations of exotic species. The face of the forests was changing to suit the needs of the PPI. In several states, SF was enthusiastically supported by international agencies. Kerala, however, did not show much enthusiasm for Social Forestry. It could not escape from its obligation to set up its FDC. The politically sensitive Keralites continue to struggle to keep their forests intact—the intellectual wing is the Kerala Sahitya Parishad which is keenly concerned with forests and forest dwellers' rights.

Summary and Conclusion

The political economy of the PPI and the political economy of forest management leave no doubt that PPI was not invited by the Kerala state government in order to generate employment, but was imposed on Kerala. Scholars and the foresters concerned who have passionately pleaded for the sole rights of the bamboo artisans to the bamboo forests, have failed to note the disjunction in the history of the basket-makers and the mat-weavers. In a market society, do they have free access to the raw materials or free access to the market? Have the props helped them to improve their quality of life? Are they at all free from exploitation? The facts show that the wages of those covered by protective structure fall far below the poverty level.

In their compassion for the deprived artisans, the scholars have tipped the raw material scale against a socially useful commodity, the newsprint produced by a well-run public sector unit. The grave danger that this unit faces because of the government's lack of concern for its own unit and the new open-door policy which has led to 'dumping' of NP by Western manufacturers has been discussed in some detail. It is difficult to visualize rounded development without newsprint, or for that matter socially useful writing and printing paper. Bamboos are one of the raw materials in India which are an ecologically worthwhile species that can be propagated to meet industrial requirements. There are also many other potential raw material species like agave or hemp which can be promoted, so that in the process location-specific, mixed crops of local species are promoted.

The current felling practices, whether it be for the survival of bamboo artisans or to produce newsprint, are damaging to the reed forest and call for the urgent need to train harvesters under careful supervision. After years of tight squeeze, the Forest Department pleads that it has scarce resources for such basic tasks. The Forest Department has acknowledged its grave error in permitting itself to have introduced ecologically damaging eucalyptus plantations into its precious forests.

Recommendations

1. The State Forest Department must be de-linked from the centre. The centre's pressures should cease.
2. It is the foresters' desire that forestry should have its own budget and quote the railways as having set a precedent. The foresters' wish should be respected.
3. All available pulpable material should be made available to HNL for it to produce socially useful newsprint.
4. It is imperative that superfluous commodities like rayon should not be allowed to be produced from pulpwood.
5. Even if HNL and FAO have certified the relative superiority of *E. grandis* over *E. tereticornis*, no new plantation should be started. Clear-felled eucalyptus plantations should be given an opportunity, perhaps be coaxed to revert to their original forests. There is a precedent in the high elevation grasslands in Kerala.
6. Bamboos and reeds must be harvested with care and felling cycles increased. New plantations of these eco-friendly species be encouraged if necessary.
7. It is imperative that FAO's and KFRI's scientific findings be shared with both the Forest Departments as well as with PPI.
8. The bamboo and reed artisan question must be viewed not at the narrow state level but at the national level and alternate sources for their development be found.
9. GOI must realize that Sustainable Development (SD) of forests is a fraud. A former director of forestry at FAO had pointed out that SD is not workable in an inequitous world and India is still a part of the inequitous Third World.
10. Thorough investigations must be carried out to identify the planters who have illegally encroached into the forest area and punishment meted out (note that the planters are politically powerful). Compensation must also be claimed to bring the land back into becoming mixed forests.

REFERENCES

Ajit Kumar 1985, 'The Impact of the Working of Kerala State Bamboo Corporation in Development of Bamboo Industry in Kerala', unpublished Ph.D. thesis of the University of Cochin.

Anon. 1976, 'Newsprint Plant by End of Next Year', *The Times of India*, 13 December.

——— 1977, 'HNL—Paper Giant in the Making', *Economic Times*, 26 May.

——— 1983, 'CBI Probing Vellor Newsprint Misdeeds', *Financial Express*, 17 October.

——— 1983, 'Poor Planning Bane of Vellor Newsprint', *Financial Express*, 4 December.

——— 1983, 'Vellor Newsprint Notches Several Firsts', *Economic Times*, 13 May.

——— 1985, 'Hindustan Newsprint Limited', *Economic Times*, 21 March.

——— 1989, 'HPC on Recovery', *Financial Express*, 9 November.

——— 1989, 'HPC Won't Seek Government Funds as Performance Improves', *Business Standard*, 5 December.

——— 1991, 'Teak Plantations of Kerala—International Teak Symposium', Kerala Forest Department, *Forests of India*, vol. 1.

——— 1992, *Review of Newsprint Allocation Policy, Newsprint Import Policy,* a letter to the Prime Minister dated 17 February 1994.

——— 1993, *Journal of Industry & Commerce*, vol. XXX, no. 5, 2-15.

——— 1993, 'NEPA Newsprint Formula Unfair', *Financial Express*, 13 April.

Basha, Chand S. 1990, 'Proc. MAB Regional Training Workshop'—*Tropical Forest Ecosystem Conservation Development S & SE Asia,* 1990: 132-43.

——— 1991, *Ochlandra (Bamboo Reed)—A Vanishing Asset of the Forests in Kerala,* S. India KFRI, Peechi (Mimeo).

Bourdillon, T.F. 1892, *Report on the Forests of Travancore.*

Chudamannil, M. 1986, 'Pattern of Investment in Forestry and its Implications', M.Phil Dissertation, CDS, Jawaharlal Nehru University.

——— 1990, Bamboos: Current Research—*Proceedings of International Bamboo Workshop*, Cochin.

FAO 1968, *Forests and People* (Henry Beresford—Peirse).

——— 1984, *Intensive Multiple-Use Forest Management in Kerala.* Forestry Paper 53 (prepared by C.T.S. Nair).

——— 1989, *Intensive Multiple-Use Forest Management in the Tropics*, Forestry, Paper 55 (prepared by A.J. Leslie).

——— 1994, *Readings in Sustainable Forest Management,* Forestry Paper 122—Contains article by A.J. Leslie (Sustainable Development of Tropical Moist Forests).

GOI 1993, *Estimates Committee Report—1992/93, Tenth Lok Sabha,* Ministry of Information and Broadcasting, New Delhi.

Government of Kerala 1990, *Forest Statistics of Kerala*, Tiruvanthapuram.

Hindustan Newsprint Limited, *Annual Reports: 1985-6, 1989-90, 1991-2, 1992-3.*

HNL 1993, *Historical Background*, HNL Publication.

Innes, C.A. 1908, *Madras District Gazetteer, Malabar*, Madras.

Jayal, J.D., U. Schettli and Chatrapathi Singh 1985, *Crisis in Forest Resource Management, Environment Cries and Response*, Natraj Publishers, Dehradun.

Jayaraman, K. and C.N. Krishnankutty 1990, *A Data Bank for Forestry Sector in Kerala*—KFRI Research Report, no. 66, Peechi.

Kallarackal, I. and C.K. Soman 1993, *Water Use of Selected Indigenous and Exotic Trees*, KFRI, Peechi.

Karunakaran, C. 1983, *A Perspective Plan for the Period 1982-83 to 1995-96 on Demand vs Import of Raw Material from Forests in Kerala State (Draft)*—Kerala Forest Department, Trivandrum.

Khan, Waheed 1971, Determination of Royalty on Pulpwood, letter to the editor, *Indian Forester*, pp. 558-61, vol. XCVI, no. 9, Dehradun.

Kothari, P.S. 1971, 'Royalty on Pulpwood', *Indian Forester*, vol. XCVII, no. 9, Dehradun.

Krishnankutty, C.N. 1991, 'Market Study of Bamboo and Reed in Kerala', in *Silviculture, Management and Utilization of Bamboo Resources in Kerala*, Final Technical Report, Kerala Forest Research Institute, Peechi.

Lal, J.B. 1985, *India's Forests: Myth and Reality*, Natraj Publication, Dehradun.

Menon, Sreedharan, A. 1962, *Kerala District Gazetteer, Trivandrum*, Government Press, Trivandrum.

——— 1962, *Kerala District Gazetteer, Trichur*, Government Press, Trivandrum.

——— 1964, *Kerala District Gazetteer, Quilon*, Government Press, Trivandrum.

——— 1965, *Kerala District Gazetteer, Ernakulam*, Government Press, Trivandrum.

——— 1962, *Kerala District Gazetteer, Kozhikode*, Government Press, Trivandrum.

Mukundan, G. and Meher Singh, undated, *Socio-Economic Aspects of Forest Eco-Systems in Kerala*, Mimeo., Forest Department, Kerala.

Nag, K. 1996, *The Times of India*, 21 April.

Nair, C.T.S. 1986, *Bamboo and Reed Workers of Kerala*, FAO, Rome.

Pascal, P. 1988, *Wet Evergreen Forests of the Western Ghats of India*, Institute Francais, Pondicherry.

Ponameshwar, Iyer S. 1974, *A Working Plan for Kozhikode Forest Division, 1964-5 to 1973-4.*

Rao, Y.A. 1989, *The Paper Industry in India—Status & Prospects*, OUP, New Delhi.

Sagariya, K.P. 1971, 'Royalty on Pulpwood', letter to the Editor, *Indian Forester*, vol. XCVII, no. 9, Dehradun.

Seethalakshmi, K.K., T. Surendran and T. Somen 1988, 'Vegetative Propagation of Ochlandra Travancoria and O. Scupta by Culm Cuttings',

Current Research Proceedings of International Bamboo Workshop, J.V.R. Rao, R. Gnanaharan, C.B. Shastri (eds.), KFRI, Peechi, 1990.

Seethalakshmi, K.K. and C.K. Soman 1992, 'Storage of Bamboo Seeds', KFRI, *Information Bulletin*, no. 12, BIC Series, no. 2, Peechi.

Shankaran, K.V. 1992, 'Litter Dynamics', *Microbial Association and Soil Studies in Acacia Auriculiformis Plantations Project,* KFRI 126/89, Peechi.

Sridharan, B.T. 1994, Indian Newsprint Manufacturing Association, letter to the Union Minister for Information and Broadcasting, dated 6 January 1994.

Stebbing, E.P. 1923, *Forests of India,* vol. II, John Lane Bodhey Head Ltd., London.

——— 1926, *Forests of India*, vol. III, John Lane Bodhey Head Ltd., London.

Suthon, A. 1988, 'A Simple and Cheap Method of Bamboo Preservation', *Current Research Proceedings of International Bamboo Workshop,* J.V.R. Rao, R. Gnanaharan, C.B. Shastri (eds.), KFRI, Peechi, 1990.

Venkatesh, C.S., K.K. Seethalakshmi and T. Surendran 1982, 'A Simple Technique of Bamboo Propagation', *Proceedings of the International Workshop on Special Problems in Physiological Investigation of Tree Crops*—Rubber Research Institute, Kottayam.

Vishwanathan, T.P. 1982, 'Kerala Forest—A Diminishing Asset', *Vanarasmi*, November.

CHAPTER 8

Tamil Nadu

Whither Tamil Nadu

The once sprawling
    Madras has shrunk,
and new Tamil Nadu
    has risen.
It needs paper,
    It needs viscose to grow.
To grow
    It needs capital,
    but not Marwari capital;
No more
    hateful northerners
Let in the southerners,
Never mind
    if he is a Brahmin,
    An Italian is not a northerner.
Give them bamboo and wattle
Let us cultivate
    stately eucalyptus
    even if they are spindly.
Of course
    there's our hero
The darling
    of the milllions,
    And
    his gift to the poor
Was to
    let them carry head loads
    To be
    transported away in truck loads,
For,
    The omnivorous paper and newsprint
    loves firewood for its pulp.

Never mind the forests,
    Yes, but where are the forests?
    Unless you mean the *sholas.*
But the Brahmin found his answers
When,
    he chanted the R.D. Mantras,
    the effluents wove black magic,
    and the dry land turned to gold.
Yet,
    the Brahmin wanted more.
Said he,
    Will you walk into my parlour?

Delighted, said the spider,
    The dyslexic Black Widow spider.
And it quickly wove its web,
    The sickly sweet web
which
    trapped the little Brahmin,
who
    trapped the Gounder woman,
And
    this trapped mother-earth.
Help!
    Wailed the omniscient,
Too late, sighed the mother

M.S.

SECTION 1

Colonial Intervention: Forests of Madras Presidency

Natural History

The former presidency of Madras had forests extendmg into Orissa, Andhra Pradesh, Mysore and Kerala. In 1956, with the reorganization of the states, Tamil Nadu lost many of these valuable forest areas.

Early accounts from Stebbing's writings reveal that a great part of the state that is Tamil Nadu today was thickly wooded. The frequent wars between the local kings—Cheras, Cholas and Pandiyas, and the Carnatic wars saw many of the forests decimated (FRI, 1961). They had no chance to recover. Trade and administrative intervention were responsible for further damage.

Stebbings (1921), says that many decades after the British took over, no schemes were launched to work forest areas. Sandalwood, ebony, teak and rosewood were exploited mercilessly by the traders and the British. In order to curb this practice a commission was set up in 1800 to regulate the felling of young teak. 1805 saw a committee being set up to monitor supply of timber to the British navy. Watson, a Police Official, who was appointed as conservator in 1806, extracted maximum amounts of timber; in 1823, however, this post was abolished.

In the mid-nineteenth century, owing to the shortage of timber supplies, a timber contractor, Williams, turned his attention to the Coimbatore forests. He applied to the collector of Coimbatore to work the forests of Anaimalai (the elephant hills) on the condition that all timber cut should be exclusively for public service.

The Government of Madras approved and Wroughton was ordered to start the work. He visited Anaimalai in 1844, but found that there would not be enough water in the streams to float out the teak till the south-west monsoon arrived in May-June. Again in 1847, Capt. F.C. Cotton submitted a report on the teak of Anaimalai to the government and asked for the services of an officer to explore it. Major-General J. Michael was appointed in June 1848. In 1849, Cotton submitted a report on the operations of felling and plan for making a road across the hills to the settlement of Colongode and Cochin boundaries. At some point of time, timber was even rolled down the mountains in the Anaimalais to the plains. A chute was constructed to slide the timber down to the plains, hence the name Top Slip (Range Officer, Anaimalai, personal communication).

Earlier when Dr. Gibson was appointed Conservator of Forests in Bombay in 1847, he was consulted in order to straighten out some of the tangles that the Madras Forest administration was facing. The work of the administration of that period was to satisfy the complaints and demands of the lessees of the forests whose claims to private ownership had been assumed, without enquiry, to be legally sound. These persons had been allotted a definite hold on large tracts of forests. Even before all this, in 1843, Black, the Collector of Madras, issued a proclamation that timbers like teak, blackwood, jack, sandal should be preserved. This proclamation went unheeded.

In 1856, the Madras government appointed Dr. Cleghorn, a surgeon, as Conservator of Forests. He is credited with introducing eucalyptus in the *sholas* of the Nilgiris in order to ensure a supply of fuel for the troops. The first plantation was commenced on a 600 acres site near Wellington. By 1858 another 8 acres of *Acacia lophanta* had been planted. Again, in the same year, the Collector of Coimbatore, E.B. Thomas, planted 8,000 Australian trees of various species around Ooty.

Why Firewood?

It is reported that Cleghorn raised the issue of raising firewood plantations in order to check the use of cow-dung. 1860 also saw a series of rules introduced in order to prevent local people from entering the *sholas* to collect firewood (Stebbing 1922). The Governor's minutes mention that some arrangements 'may be made' to grow casuarina around the wastelands of each village. The British felt that casuarina grew easily on the Madras coast, on the Mysore Plateau and in the Nilgiris. Planting of casuarina had already begun in Madras as early as in 1866.

However, despatch register no. 2 dated 12 February 1867 from the Governor of Madras to the Secretary of State contained information on the steps taken to increase the general supply of fuel in the country, that the largest of the jungles was in Salem district and that 12 sq. miles had already been reserved for firewood extraction. Firewood was regarded by the government to be vital for the troops stationed in the hills and for the railways. The railways at that time used wood as fuel, emerging as more than a serious competitor to the villagers. The ever-expanding demands of the railways led the Board of Revenue to instruct the Collector of Coimbatore to mark off and reserve lands. In June 1866, Lt. Walker was engaged in the task of reserving 5,000 acres for this purpose at the extreme western limit of Coimbatore district and near the railway lines.

Dr. Cleghorn, in consultation with his officers, decided to form forest plantations to meet the fuel demand of the railways. These plantations were to be located near Erode, Bangalore, Cuddapah and Arakonam, with either the railways or the government taking up the task.

There was trouble in carrying this out, since the collectors did not realize the need for forest conservancy and supported the Forest Department only half-heartedly. The Collectors were supported by the Revenue Board. The government felt threatened, because, should

England enter into a naval war, coal and coke supply from England would stop. In India, the nearest coalfields were in the north.

The Secretary of State vetoed the plan for the railways to take up their own plantations. The plan to form plantations at Cauverypooram came to nothing. In a letter written in 1869 by one of the consulting engineers, the reasons become clear. He pointed out that 'there are inexhaustible forests' near the Cauvery river above Erode and all that would be needed was a tramway to bring the wood to the river. He also marked out other forests nearby which could be worked, like the dense *shola* forests on the Bolampatty hills, the Palghat hills and the Anaimalais. He went a step further and stated that the woods in these forests were of no use except as fuel. In the same letter a report on the plantation activities is given. About 600 acres near Gooty and Cuddapah were planted. Also, 30,000 young trees of casuarina were planted out near Tirunelveli (Stebbings 1924).

Meanwliile, the Madras Railway Company placed a request with the Secretary of State asking him to increase the fuel rates. A conference was called of consulting engineers, conservators and the agent and manager of the company. As a result, forest tracts were demarcated in Cuddapah, North Arcot, Salem and Coimbatore to an extent of 1,7,000 acres to provide fuelwood to the railways.

The progress of forest conservancy in Madras Presidency was slow and difficult owing to the unfortunate attitude taken by the revenue officials, comments Stebbings (Stebbing 1923). The Indian Forest Act of 1865 and the subsequent Act of 1878 (for the constitution of reserved and protected forests), was not extended to Madras due to stout opposition from the Revenue Board (FRI, 1961). Stebbing also adds that the Collectors of the districts were placed in a difficult position, more so since the Board of Revenue was unable to make up its mind to order proper surveys of forests. Some progress had been made with fuelwood plantation work in Nilgiris, Wellington, Wynaad, Madurai, Golconda and casuarinas on the coast. In 1881, Brandis visited Madras and a Forest Act for Madras was constituted in 1882. He said:

> The advantages expected to be secured by forest conservancy in Madras Presidency as in other provinces of the Indian Empire, were direct and indirect. The direct advantages comprise the provision of a permanent and sufficient supply of timber wood and other forest produce for local use and for railways. The indirect advantage would influence climate, underground and surface drainage. (Brandis 1883: 4)

By 1900, the state of Madras had 20,000 acres of plantations of *eucalyptus*, *Acacia mearnsii*, *Acacia decurrens* (Wattle), and casuarina on the east coast. Eucalyptus was introduced in Ooty, viz., *Eucalyptus globulus*, the blue gum, in 1843 by Captain Cotton, while the first government plantation was initiated by Hutchinson in 1862.

In 1914 certain reserve forests near villages were converted into forest panchayats. In 1923 a committee was set up to demark major and minor forest areas, the former being reserved forests and the latter to be given over for village needs. Forest panchayats were destructive to the forests and the folly was realized only as late as 1949 when the 'Panchayat Forest' system was eliminated.

In 1923 a forest engineer was appointed to extract timber. With World War II, large-scale exploitation outside working plan prescriptions was carried out. Even burning of wood for charcoal was permitted within the forests.

At the close of the century, only 19,649 sq. miles were reserved forests. As of 1960, 19,000 sq. kms were reserved forests (FRI, 1961).

SECTION 2

Forests of Tamil Nadu

The catchment of the first PPI unit in the state shifted from the north and its immediate surrounding areas to Coimbatore and the Nilgiris, as the bamboo yield depleted and vanished. The demand began to rise as two new units were set up. The demand was attempted to be fulfilled by artificial forests—plantations of exotic species grown in the forest area as well as in the plains. The scenario therefore is rather broad-based and will therefore be dealt with at other levels.

An account of the physiography of the state of Tamil Nadu, as given by the old and new Gazetteers is first presented here. The relief features of the districts of Nilgiris and Coimbatore will be dealt with in detail as they are the chief catchment areas today.

Relief—Tamil Nadu

The Western Ghats run almost parallel along the whole length of the western coast, as do the Eastern Ghats on the east coast. These two ranges eventually meet, to form the Nilgiri hills and the Anaimalais. The spurs of the Western Ghats in Tamil Nadu is the Anaimalai range, a continuation of which extends into Madurai district, the Palani range. Several isolated blocks of hills like the Shevaroys in Salem, the Pachamalais and the Kollaimalais in Salem and Trichy, the Javadi hills in the north and south Arcot stand isolated from the continuous ghats.

Thus the state of Tamil Nadu consists of a narrowing stretch of land between the Western Ghats and the Indian Ocean. Within this an elevated tract lies between the Western and Eastern Ghats.

Drainage: The whole trend of the drainage is from west to east into the Bay of Bengal. The Cauvery remains the largest river rising in the Western Ghats passing eastward into the Bay of Bengal. Smaller rivers like the Pennar, the Ponnaiyar, the Tambiraparni follow the same direction. In the early phase of their courses these rivers drain the country as they run into deep beds, but nearer the coast, dams across these rivers have utilized the waters for irrigation.

Rainfall: Climatic conditions of the state are varied. While the coastal regions record the highest temperatures, the hill districts record the lowest. The chief rain-bearing current is the south-west monsoon, from May to September, which blows in from the Indian Ocean. But this is unable to pass over the Western Ghats and little rainfall is received. As the force of the south-west monsoon in north-eastern India dies away, the current curves, driving inland from the Bay of Bengal during October to December, bringing with it the north-east monsoon. The rainfall is the heaviest along the coast but as the clouds drift inward they gradually give less rainfall. The central and southern lands get only moderate rainfall.

Soils: The deltas and river margins have rich alluvial soils. There are large tract of black cotton soil too. But the soils of most areas are gravelly, shallow and sandy, overlying an impenetrable subsoil of gravel and rock. Even the black soil is inferior, saline and shallow. Nitrogen is wanting, humus is low, iron is high; report the *Imperial Gazetteers*, Madras (1908).

Forests: The most important forests are in the south and west on the Western Ghats and the Anaimalais and in the Nilgiri hills, say the Gazetteers of 1908. These were evergreen forests with timber species, but, being inaccessible were left undisturbed. Portions of the Eastern Ghats running through Coimbatore and the row of hills in Salem and north Arcot, viz., the Shevaroys, and the Javadi hills, were less wooded. These forests had very little timber species except for sandalwood. The early Gazetteer does not mention the prevalence of bamboo forests.

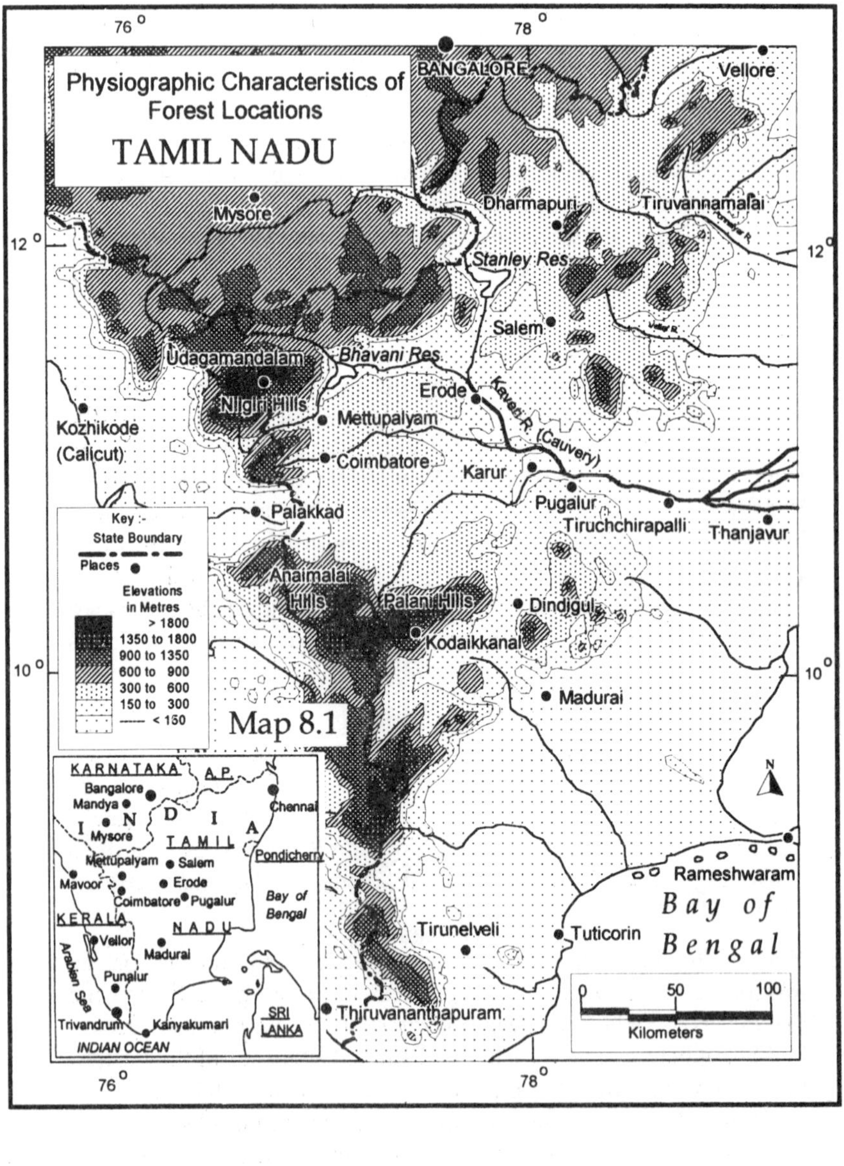
Physiographic Characteristics of Forest Locations
TAMIL NADU
Map 8.1
BANGALORE
Vellore
Mysore
Dharmapuri
Tiruvannamalai
Stanley Res.
Salem
Udagamandalam
Bhavani Res.
Nilgiri Hills
Erode
Kaveri R. (Cauvery)
Mettupalyam
Kozhikode (Calicut)
Coimbatore
Karur
Palakkad
Pugalur
Tiruchchirapalli
Thanjavur
Anaimalai Hills
Palani Hills
Dindigul
Kodaikkanal
Madurai
Rameshwaram
Tirunelveli
Tuticorin
Bay of Bengal
Thiruvananthapuram
Key :-
State Boundary
Places
Elevations in Metres
> 1800
1350 to 1800
900 to 1350
600 to 900
300 to 600
150 to 300
< 150
0 50 100
Kilometers
76°
78°
12°
10°
KARNATAKA
A.P.
Bangalore
Mandya
Mysore
I N D I A
Chennai
TAMIL NADU
Pondicherry
Mettupalyam
Salem
Mavoor
Erode
Coimbatore
Pugalur
Bay of Bengal
KERALA
Vellor
Madurai
Arabian Sea
Punalur
Trivandrum
Kanyakumari
SRI LANKA
INDIAN OCEAN

Physiography of Coimbatore

Relief: The district of Coimbtore is traversed by hill ranges in the north, in the north-west and south by mountains of the Western Ghats (the Nilgiris in the north and the Anaimalai in the south). The Cauvery river forms the eastern boundary. The rest of the district is a plain, sloping gradually from the west to the east. This is a largely undulating plain, except in the black cotton soil tracts of Udumalpet, Palladam and Coimbatore taluks. A few hills of slight elevation are found in Erode and Dharapuram taluks.

The most dominant mountains are the Anaimalais which join the Palani hills of Madurai. The lower ranges to the west comprise the teak forests and all the hills here are covered with forests, while on the higher ranges, there are open grassy hills and valleys with *shola* forests. The other hills are the Vellangiris, the Sholakarai and the Bolampatty.

Drainage: The chief rivers of the district are the Cauvery and its tributaries, the Bhavani, the Noyyal and the Amravathi are the other main rivers.

Minerals: The district does not have significant mineral wealth. The soils of the district are chiefly red sand and gravel with a moderate area of red loam and black loam or sometimes black clay. In the uplands the soils are stony, gravelly and extremely thin, but at the bottom of the valleys, the soil is of moderate depth and fairly rich.

Rainfall: The district is exceptionally dry and rainfall is scanty. Maximum rain is afforded by the north-east monsoons. Rainfall varies from 23.78 to 40.12 inch in north Coimbatore and from 20.89 to 209 to 296 inch in south Coimbatore. The temperature varies from 75 °F to 105 °F during the day and 65 °F to 85 °F during the night. During the south-west monsoon, strong winds from the west affect the vegetation of north Coimbatore, causing the trees to have asymmetric shapes.

Relief of Nilgiris

Relief: The Nilgiris district is bounded on the north by the former state of Mysore, on the east and south-east by Coimbatore district, and on the west and south-west by Malabar. It has two well marked divisions, the high steep plateau formed by the junction of the Eastern and

Western Ghats, and a lower area forming part of the Malabar Wynaad. The plateau is upheld by hills of grassy downs divided by narrow valleys. In the wrinkles of the hills nestle the *sholas*. The lower area of Wynaad or Gudalur is level and covered with thick forests (*Imperial Gazetteer*, 1908).

Drainage: To the extreme west of the district, Pandi a tributary of the Beypore river, drains towards Malabar. The Bhavani runs east under the southern wall of the plateau into a deep valley and is joined in the east by the Moyar. The Moyar rises in the Makurti peak (of the Kundah range which forms the south-western edge of the plateau) and is known as the Pykara. It drops into the lower country in the north-west to form the Moyar.

Temperature: The climate is cool. The western parts of the district receive the bulk of the south-west monsoon rains, while much of the plateau from December to April is almost rain-free. Gudalur gets the heaviest rainfall.

Soil: On the plateau the soil is dense and of a peaty nature. Soils of the rich black loam type, the brown type and the inferior yellow and red types are also found. Gudalur has marshy alluvial flats along the low hills and receives heavy rainfall. On the plateau only dry crops are raised (*Imperial Gazetteer*, 1908).

Forests: The hills have forests of the evergreen and the deciduous type on the slopes. The *sholas* on the plateaus were left untouched as the trees had no commercial value. The first plantations of Australian acacia and eucalyptus were planted to meet the firewood demand.

Status of the Forests of the Nilgiris and Coimbatore

These two forest circles are the catchment areas of three PPI units.

Districts: Given below is an account of the status of vegetation in the districts of Nilgiris and Coimbatore. The picture of the vegetation, as recorded by Gazetteers of the early twentieth century is coupled with accounts of studies made by scientists before the industries were set up and the picture of the vegetation, as recorded by the team's own observations, after the industries were set up. The changing face of the forests may or may not be due to the paper industry alone. The catchment area of the two paper mills and viscose rayon are, however, located in these districts.

The plantations created specifically for PPI have indeed changed the face of the forests and are discussed under head 'Augmentation of Forest Resources'. There was no bamboo felling for us to make our observations on felling practices by the mills.

Past Status

The vegetation of the Nilgiris was classified into four as:

The Deciduous Forests: On the slopes were species of *Erythrina, Butea, Dalbergia* and *Terminalia*. In these forests a large number of shrubs, sedges and grasses were found growing between the valuable timber trees.

The Moist Evergreen Forests: These were found at 3,000 ft and 4,000 ft elevations. Many orchids, ferns and mosses were common to these forests. Characteristic of these forests was the *Strobilanthes* shrub. Three species of reed/bamboo flourished here—*Ochlandra rheedei, Oxytenanthera thwaitesii* and *Teinostarchym wightii.*

Above 4,000 ft these forests decreased in size and pass into *sholas*. The timber of the moist evergreen forests was not of good quality except for *Mesua ferrea* and ebony.

The Sholas: The trees here were of a different *genera* and grew to heights of not more than 70 ft. They are composed of *Eurya japonica, Gordonia obtusa, Michalea nilgarica, Ternstromia japonica* and various *Lauracea* species (Lal 1985: 54) but orchids are not found here. Ferns and mosses grew in plenty.

The Grasslands: Many species of short, coarse species of grass which die in the frost and come up in the spring are found in the plateaus. Trees like the *Rhododendron, Salix, Celtis*, etc., were sparsely scattered.

The accounts of the early Gazetteers highlight the predominance of introduced species like eucalyptus and acacia which altered the landscape. The gloomy foliage did not beautify the area; also, the Forest Departments were putting in many trees of *Eucalyptus globulus, Acacia melanoxylon* and *Wattle* (Francis 1908). As early as 1908 the forest cover was confined to the top of the plateau and the upper edges. The Gazetteers also point out that there may have been many more *sholas*, but their spread had been reduced due to grazing pressures by cattle belonging to the local Badaga population. While the locals did not cut down trees initiating any disturbance to the ecosystem, the onslaught of the European settlers in no way contributed to the regeneration of the system.

The district of Coimbatore had forests mainly in the Anaimalais. The vegetation of North Coimbatore varies from that of South and is classified separately.

North Coimbatore

Southern Tropical Thorn Forests: Found at the foot of the hills, plains and gentle slopes having low rainfall as in Udumalpet and Sathyamangalam areas. *Acacia spp.* and *Commiphora caudata* are dominant.

Smaller trees like *Randia spp.* and shrubs like *Dodonaea viscosa*, *Cassia auriculata*, etc., are found here, while grass growth was sparse, and the soil exposed and bare.

Southern Tropical Dry Deciduous: Found along Yekkattur Ghats, Onnaithittu, Gadessal, Talamalai and Bargur.

Anogeissus latifolia, *Pterocarpus marsupium*, *Dalbergia latifolia* and rarely teak, *Bambusa arundinacea* and *Dendrocalamus strictus*, the latter being dominant, were also found here. At places like Guttiyalattur and S. Bargur, *Dendrocalamus* formed pure stands. These forests are highly denuded.

South Coimbatore

Southern Wet Temperate Forests: Found at altitudes beyond 1,666 m in the Punachi range as patches amidst the grasslands. They are confined to sheltered folds and depressions where there is adequate soil moisture. Many short-boled, straggling, closed evergreen trees of low height are found. Common species include *Syzygium spp.*, *Michelia spp.*, *Viburnums*, *Rhododendron*. Dense undergrowth of shrubs and herbs was characteristic, being interspersed with vast stretches of grasslands.

Tropical Wet Evergreen Forests: Here there is sparse undergrowth and heavy leaf litter, found in regions between the elevation of 2,000 ft to 5,000 ft. The forest was multi-storeyed with tall emergent dominants. Principals here were species like *Mesua ferrea*, *Hopea parviflora*, *Artocarpus spp.*, *Diospyros spp.*, *Mangifera indica*, *Drypetes*, *Mallotes spp.*, *Hydnocarpus spp.*, etc.

Southern Moist Deciduous Forests: Confined to altitudes of 1,500 ft. to 3,500 ft. Trees were long-boled, *Tectona grandis*, *Terminalia spp.*, *Grewia tiliaefolia*, etc., being the dominant species. In open canopy areas, pure colonies of *B. arundinacea* were found.

Southern Tropical Dry Deciduous Forests: These were found along the foothills of the Pollachi range. These forests were similar to those of the northern part of the district.

Southern Tropical Thorn Forests: These were confined to the eastern portion of the Pollachi range over Udumalpet range, again similar to the ones found in north Coimbatore.

Present Status of the Forests

A large portion of the two districts is covered by forests. The forests on the hills in the eastern foot of the Nilgiris do not produce very good timber. The higher ranges of the Anaimalai contain much valuable timber, the lower ranges being the teak zones. Bamboo is now practically absent in the plains, though there are many rivers and tanks where it can be grown to advantage.

The forests have been divided into the North Coimbatore Forest Circle (Erode, Bhavani and Gobichettipalayam) and the South Coimbatore Forests Circle (Avinashi, Coimbatore, Pollachi, Udumalpet and Dharapuram). The forests of the district cover 2,272 sq. miles of Reserve Forests and 24 sq. miles of private forests.

In the north, at the foot of the Bargur hills, submontane forests of degraded scrub, thorny species and almost xerophytic types are found, and on the higher reaches on the plateaus, forests containing sandal. Also found are both superior and inferior deciduous type of forests. The former is of the teak-bearing kind, while all the latter are close to villages on the plains. A few patches of *shola* forests are found on the Bargur range. The north Coimbatore plateau forests are of the mixed deciduous species of poor growth and height. The forests on the slopes are rocky, ill-protected and unproductive.

On the higher ranges, evergreen and semi-evergreen type of forests are found. The bamboo forests grow most densely on the outer slopes of the north Coimbatore plateau. The main species are *Dendrocalamus strictus* and *Bambusa arundinaceae*, also found along river banks. The main bamboo areas are Vadaparai, Gundri plateau, Gutialattur and the Hassanur basin in the Satyamangalam range.

In 1986, all the bamboo flowered and was removed and given to the industry. The sanctuaries remain the last frontier for the pristine vegetation of Tamil Nadu. As seen in the Anaimalai, good dense growth of both *B. arundinacea* and *D. strictus* can be observed. The forests have begun to resurrect, but those which were lost, cannot be recreated.

The blue mountains, the Nilgiris, are now blue with blue gum (*Eucalyptus globulus*) and plantations of acacia, wattle, pines, etc. Very few patches of grasslands and *sholas* now exist.

PPI and their Catchment Areas

The three PPI units of Tamil Nadu, viz., Seshasayee Paper Boards Ltd. (SPB); South India Viscose Ltd. (SIV) and Tamil Nadu Newsprint Ltd. (TNPL) and their catchment areas will be analysed in detail.

While bamboo was the principal raw material when SPB was set up in the 1960s, now it is no longer so. SPB was allotted areas in south and west Vellore, north and south Salem, Tiruchirapalli, north and south Coimbatore, Ooty and Gudalur; all of which were once rich in bamboo and have now ceased to be so.

Since the early 1970s bamboos were replaced as a raw material by eucalytpus. *Eucalyptus globulus* planted in the Nilgiris is allotted to South India Viscose, while *Eucalyptus grandis* and *E. hybrid*, which were planted in the plains at the Anaimalai foothills as well as the Nilgiris are allotted to SPB and TNPL. Yet another industry which is dependant on plantations is the tanning industry which extracts tanning from the bark of the wattle trees (*Acacia mearnsii* and *Acacia decurrens*). Wattle is yet another plantation that has covered the south-western hill slopes of the Nilgiris. The tan industry requires only the bark, the debarked wood being allotted to the PPI units. (Details of quantity supplied are given in the section on raw materials.)

When the industry started out it had access to the forests of both the Eastern and Western Ghats. Having exhausted almost all the bamboos in the region, the catchment area has become restricted. The industry is now allotted plantation species and only dead and dry bamboos. As of today, the chief catchment areas are located in the districts of Nilgiris and Coimbatore. Table 8.1 indicates the raw material supplied by the FD of Tamil Nadu to the three PPI units.

Augmenting Forest Resources

The principal agencies involved in meeting the raw material requirements of the three pulp and paper mills of Tamil Nadu are the Forest Department (FD) and the Tamil Nadu Forest Development Corporation (TAFCORN).

Company sources at Seshasayee Paper Mills said they were using bamboo (when supplied by the FD), wood, bagasse, waste paper and imported pulp as their raw materials. The wood was mainly *E. tereticornis* or hybrid, *E. grandis*, *Casuarina*, *Pinus petula*, *E. globulus*, *Erythrina indica*, *Erythrina supersa*, *Subabul*, *Acacia nilotica*, *Cashew*, *Lantana* and pineapple leaves. While the eucalyptus was being supplied through FD and TAFCORN, the casuarina and other woods mentioned, were bought from private farmers. While Tamil Nadu Newsprint Limited is largely bagasse-based, it also uses these woods in enormous quantity.

Role of the Forest Department (FD)

One of the major industries that the FD is catering to is the pulp industry (Table 8.1). The PCCF of Tamil Nadu was of the opinion that bad rainfall over the years and lack of good forests have resulted in a shortage of raw materials for the industry. Tamil Nadu also has a tanning resource base—wattle.

In the 1950s, with the reorganization of the states, Tamil Nadu lost many of its productive forests to other states. In 1956, the then PCCF, launched a programme to promote *E. globulus* (blue gum) and wattle at the expense of the *sholas* and the grasslands. Pulp and Paper consultants commissioned by ECAFE–FAO did make such a suggestion. The consultants had suggested 16,000 acres be planted with wattle and 10,000 acres with blue gum (FAO, 1962: 35). Within 10 to 15 years the hills were covered with exotics. A local outcry did break out but not long enough to alter the situation.

Survey of Bamboo Forests of Tamil Nadu

When Seshasayee Paper Boards (SPB) was to set up its factory at Erode, the then-PCCF was asked by the government in 1959 to conduct a survey of the bamboo resources. Outlined below are the details of the scheme. This scheme was devised to make bamboo available for a 60 tonne paper plant without affecting the rural economy. Areas that fell within the scope of the scheme were east and north Vellore, north and south Salem, Tiruchirapalli, north and south Coimbatore, Ooty and Gudalur, all within a 150 miles radius of the first PPI factory SPB, in Erode.

Only two species of bamboos were available, *B. arundinacea* and *D. strictus*. The surveyor pointed out that till then, the local needs were

TABLE 8.1: PULPBOARD SUPPLIED TO VARIOUS INDUSTRIES (1978-9 TO 1992-3)

Year	South India Viscose Limited				Seshasayee Paper & Boards Limited				Tamil Nadu Newsprint Paper Ltd.			
	Pines	*Bluegum E. globulus*	*E. grandis*	DW Wood	Pines	Bamboos	E. hybrid	D.W. Wood	*E. globulus*	E. hybrid	D.W. Wood	*E. grandis*
1978-9	—	29,753	3,974	19,865	—	26,444	35,037	—	—	—	—	—
1979-80	—	54,000	10,259	41,188	—	29,282	17,167	—	—	—	—	—
1980-1	—	57,976	6,937	25,945	—	47,006	8,054	—	—	—	—	—
1981-2	—	61,316	22,156	43,056	—	20,629	30,328	—	—	—	—	—
1982-3	—	27,746	10,153	22,859	—	23,371	30,389	—	—	—	—	—
1983-4	—	19,606	5,636	18,080	—	17,382	32,041	—	—	—	—	—
1984-5	—	58,280	22,347	52,899	—	11,713	38,640	2,999	11,999	—	—	1,792
1985-6	—	58,862	24,557	29,665	—	11.108	26,447	15,000	8,051	—	—	14,257
1986-7	—	83,731	8,968	30,345	—	12,541	30,310	11,091	2,564	—	7,258	—
1987-8	—	85,033	5,333	29,504	—	4,672	6,777	12,552	2,100	—	23,750	—
1988-9	—	71,000	6,000	41,500	—	8,000	25,000	12,000	—	—	16,500	—
1989-90	—	71,817	4,915	53,461	—	2,000	12,477	12,074	4,636 (B. gum)	—	12,004	—
1990-1	—	62,800	72,00	65,000	—	10,000	9,000	14,000	—	12,600	14,00	—
1991-2	—	83,000	5,500 (Hyb.)	32,000	10,000	4,000	3,000	16,000	—	5,500	24,000	5,500
1992-3	5,000	54,000	7,300	51,500	5,000	5,000	12,000	8,500	—	15,500	14,500	—

Source: Tamil Nadu Forest Department.

Note: DW—Debarked Wattle.

TABLE 8.2: BAMBOO YIELD FROM AREAS SET APART FOR INDUSTRY

District	Range	Yield by		Yield taken into consideration (in tonnes)	Contractors outturn (in tonnes)	Yield as per 1956 Survey (in tonnes)
		Annual Recruitment basis	Total no. of Green Culms			
Coimbatore North	Bargpr (Part)	4,589	3,227	4,589	5,378	
	Andipur (Part)	638	890	638	502	8,000
	Talamalai	412	891	—	680	
	Satyamangalam (Part)	5,927	3,833	5,927	2,182	11,000
Coimbatore South	Topslip Range	2,900	8,597	8,597	5,998	Survey not done
	Pulachi Range	7,718	3,578	3,578	n.a.	19,631
Gudalur	Madumalai Range	621	653	653	n.a.	—
	Cherambady Range	125	567	567	n.a.	—
	Sigur Range	195	870	870	n.a.	—
	Private Forests	1,616	4,532	4,532	n.a.	15,363
Salem North	All Ranges	—	—	9,143 (as per WPO)	n.a.	12,000
Salem South	Shevaroys North (Part)	1,092	3,530	3,530	5,808	5,000
Vellore West	Sathanur Range	2,998	5,298	5,298	189	7,500
				49,234		78,494

(Contd.)

TABLE 8.2 (Contd.)

Division	Range	Yield (in tonnes)	Contractors outturn (in tonnes)
Coimbatore North	Sathyamangalam (Part)	1,288.44	1,582
	Andiyur (Part)	630.15	1,585
	Bargur (Part)	693.29	1,066
	Mettupalayam (Part)	1,236.73	1,000 (approx.)
Coimbatore South	Bolampatty	715.64	850
	Pollachi (Part)	78.81	360
	Udumalpet (Part)	192.29	205
Ootacamund	Coonoor	350.30	—
Vellore East	Polur	1,798.00	2,368
	Santavasal	546.00	962
	Odugathur (Part)	23.00	32
Salem South	Anthur (Part)	2,152.00	758
	Shevaroys South	1,415.00	2,531
Vellore West	Tirupathur (Part)	547.00	691
	Alangayam (Part)	1,962.00	128
	Chenham (Part)	584.00	817

Source: A Report on the Bamboo Resources of the State (FD) by Sri John Joseph (IFS) 1959.

being met from the *patta* lands and the unreserved areas. Due to heavy extraction in the past, bamboo was not very abundant even in the 1960s. The working was not carefully supervised because of lack of adequate staff in the forest department.

Detailed accounts of available stock and estimated yield were made for each area. This estimate was 30 per cent less than that made by the 1956 survey. The contractor's outturn shows no relation to the estimates (Table 8.2).

After considerable planning it was recommended that areas for extraction be allotted to the contractor under supervision by the FD.

The bamboo areas were given over on annual leases to the company and the contractors extracted maximum amounts with little regard for the prescriptions of the working plans.

G.O. no. 13010/FR.V/77-1 dated 22 March 1977, stated that 'in view of the decreasing forest cover, felling in natural areas is to be stopped'. The Chief Conservator of forests had to assess whether a permanent ban could be imposed so that further land for teak and eucalyptus may not be made available. Also, since the farm forestry schemes had started yielding, it was not justifiable to cut forests for fuel.

In 1960 the state government launched farm forestry schemes to provide tree-cover on barren lands, village common lands, road sides, etc.

Social Forestry

In the report of the National Commission on Agriculture (1976: 45) it is suggested that Social Forestry (SF) must meet the rising domestic and export demand of industrial wood. Social Forestry was started in Tamil Nadu with aid from the Swedish International Development Authority (SIDA) in order to involve rural communities in raising wood for fuel, fodder, food, fibre, small timber. SIDA's role was mainly as the funding agency providing 70 per cent of the funds.

According to the Chief Conservator of Forests—Social Forestry CCF (SF), this scheme was launched with a view to reduce the pressure on the forests for firewood. The scheme has been a success, he said, and wholly supported by the people. Social Forestry includes areas in the Reserve Forests, buffer zones of sanctuaries and the homesteads. The CCF (SF) also pointed out that 80 per cent of the domestic fuel requiremets were being met through agricultural wastes. If this is correct it made one wonder as to who was creating the firewood demand on the forests, since the local needs were being met through agriculture residues.

It is interesting to note that while the government was effective in putting a ban on collection of firewood by contractors through the FD's 1975 ban, it could not deny the local people the right to firewood. People were permitted to collect 'headloads' of firewood from forests, with no restriction on the number of trips per person per day. We were told that on an average 1 lakh headloads of firewood go out of the forests. These are sold to contractors and taken away by the truckload.

Where was all this fuelwood going to if the people did not need it? Some said it was to meet the urban demand. How much of a demand could there be in the face of present availability of alternatives like coal, kerosene, and other petroleum products'? Any type of wood can and is being pulped to produce paper.

The industry says that it gets a mixture of hardwoods, casuarina, acacias, etc., from private sources. Which are these private sources?

These are the very species raised under the Social Forestry schemes. The mills conceded that it buys wood from Social Forestry. Does the state really need fuelwood plantations? And if so, for whom? Social Forestry since the late 1980s has been modified to incorporate the concept of bio-diversity (SIDA—*Report*).

FD's Pulpwood Supply

The FD said that it could supply 3 lakh tonnes of pulpwood whereas the industry is asking for 10 lakhs. The department has stopped raising any more eucalyptus plantations and plans to withdraw all the plantations from the hills and raise pulpwood on the plains and non-agricultural lands. As it stands today, 17.50 per cent of the state is under forest cover of which 25 per cent is degraded, 8 per cent is under production forestry, 13-14 per cent under wildlife conservation projects and the rest under conservation. The few good forests are on the Western Ghats.

On the bamboo issue, the FD revealed that after the flowering which took place in the 1960s, bamboo regeneration has not been good. There should be a stress on the artificial regeneration of bamboo since re-stocking is necessary. The age-old planting methods practised in the Thanjavur districts where bamboo was raised along the fringes of paddy fields needs to be perpetuated (Andiappan and Wilson 1963).

Tamil Nadu Forest Development Corporation—TAFCORN

TAFCORN was set up in 1974 to raise plantations to meet the industrial demand, all its three PPI units being joint-sector units. TAFCORN acquired land on lease from the FD and also had some FD plantations transferred to it for maintenance. TAFCORN jurisdiction extends only over the plains areas.

74,000 ha was transferred to TAFCORN in 1974. Every year it harvests from 5,000 ha and replants 3,000 ha. About 40,000 ha are under eucalyptus, 20,000 ha cashew, while river-beds have been planted with casuarina. There are separate corporations for tea and cashew also. Our concern here is with euclapytus.

Figure 8.1 and Table 8.3 will give an idea of the pulpwood supplied to the paper mills and the cost of raising these plantations. Tables 8.4 and 8.5 show the extent of plantations.

The annual report of the company for 1990-1 says that a Project

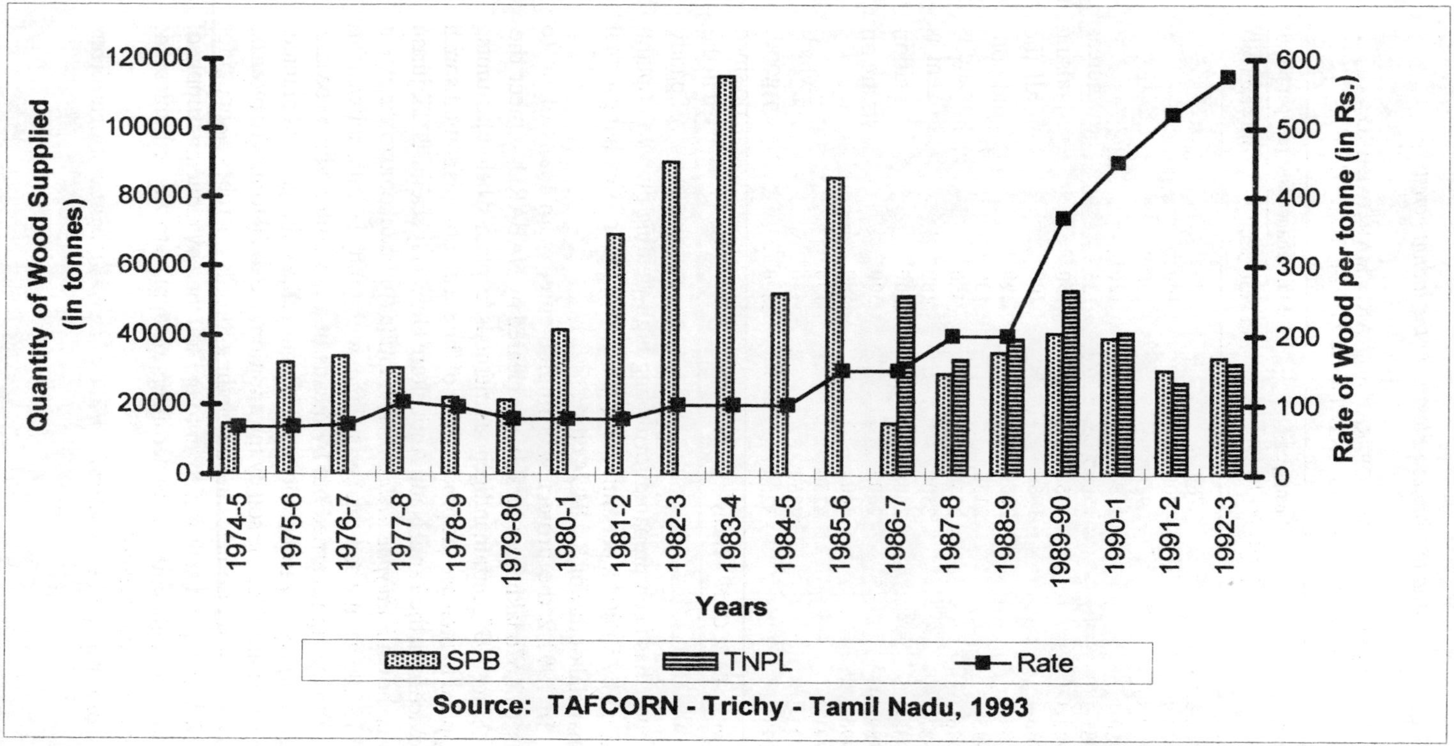

Figure 8.1: Details of Pulpwood Supplied to Paper Mills (1974-5 to 1992-3)

TABLE 8.3: COST OF RAISING AND MAINTAINING EUCALYPTUS PLANTATIONS (1977-8 TO 1993-4)

Year	Raising cost rate per ha (in Rs.)	Maintenance cost rate per ha	
		2nd year (Rs.)	3rd year (Rs.)
1977-8	780	75	—
1978-9	800	100	—
1979-80	910	120	—
1980-1	1150	120	—
1981-2	1580	125	—
1982-3	1750	150	—
1983-4	1930	200	—
1984-5	2050	250	—
1995-6	2150	400	—
1986-7	2350	450	—
1987-8	2500	480	—
1988-9	2600	500	—
1989-90	2600	500	—
1990-1	2600	500	—
1991-2	4250	350	—
1992-3	4250	550	—
1993-4	5250	2800	550

Source: TAFCORN—Trichy—Tamil Nadu.

Report has been prepared for raising high-yielding pulpwood plantations by 5,000 ha per annum for the next six years, involving a total financial outlay of Rs. 84.74 crore.

In 1991-2 the project was implemented with loans of up to Rs. 612.676 lakh from local banks cleared by NABARD.

Since 1990, certain indigenous methods (explained below) are being applied to increase the productivity of the eucalyptus plantations, with advice from the Tamil Nadu Agriculture University scientists. The team was shown *Eucalyptus tereticornis* plantations in Pudukottai to the extent of 15,000 ha; it was a total wasteland, with sandy lateritic soils.

Ploughing is carried out by means of a disc, after which seeds of *Teprocia purpurea* are broadcast. This legume fixes nitrogen and provides green manure for the soil by the addition of 5 to 20 tonnes of biomass to the soil. It also helps in soil-binding, breeds microbes essential for soil fertility, and acts as a cover to the soil thereby reducing the loss of water through radiation. Biofertilizers are applied to the seedlings in the pit and in the nursery.

Biofertilizers like *Assasparillum* and *Phospho-bacteria fix* nitrogen. They

TABLE 8.4: AREA PLANTED BY TAFCORN (1974-5 TO 1980-1

(*in ha*)

Year	Trichy				Pudukotai			
	Eucalyptus hybrid	Cashew	Casuarina	Miscella-neous	Eucalyptus hybrid	Cashew	Casuarina	Miscella-neous
1974-5	4.50	211.51	59.60	—	845.72	80.00	40.60	—
1975-6	207.12	38.20	6.00	—	1.366.91	204.07	192.37	—
1976-7	396.00	70.00	—	—	1,570.00	261.00	187.00	—
1977-8	173.71	30.00	12.72	—	1,322.41	56.52	52.72	—
1978-9	181.41	43.90	—	—	1,372.58	3.85	88.36	—
1979-80	332.01	3.20	21.08	—	1,022.45	—	—	—
1980-1	277.89	—	31.75	—	620.48	—	9.40	99.21
Total	1,572.64	396.81	131.15	—	8,120.55	605.44	570.45	99.21

(*in ha*)

Year	Cuddalore				Karaikudi		
	Eucalyptus hybrid	Cashew	Casuarina	Miscella-neous	Eucalyptus hybrid	Cashew	Casuarina
1974-5	530.00	51.00	42.00	—	489.59	325.20	135.00
1975-6	1,141.67	139.50	—	—	937.20	94.80	85.10
1976-7	897.00	21.00	—	—	1,115.00	—	32.00
1977-8	630.59	34.00	—	—	695.84	20.67	45.05
1978-9	1,010.72	30.00	—	—	927.02	10.17	19.99
1979-80	643.30	5.00	—	—	843.04	—	—
1980-1	660.01	15.00	8.18	35.00	962.81	—	—
Total	5,513.29	295.50	50.18	35.00	5,970.50	450.84	317.14

Project Officer

Kodaikanal	1977-8	1978-9	1979-80
Bluegum	46.00	92.00	—
Geranium	—	4.00	50.00

TABLE 8.5: PLANTATION UNDER FOREST DEPARTMENT

	Eucalyptus	Cashew	Casuarina	Miscellaneous	Bluegum	Geranium
Trichy	1,572.64	396.81	131.51	—	—	—
Pudukottai	8,120.55	605.44	570.45	99.21	—	—
Cuddalore	5,513.29	295.50	50.18	35.00	—	—
Karaikudi	5,970.50	450.84	317.14*	—	—	—
Kodaikanal	—	—	—	—	138.00	54.00
Total	21,176.98	1,748.59	1,068.92	134.21	138.00	54.00**

Notes: *Transferred to Forest Department.

**Transferred to Cinchona Department.

mobilize more plant nutrient from the soil, making it available for the plant. Till now this was only tried out on an experimental basis; from this year on, they plan to start it on a large scale.

The biofertilizer is mixed with the soil medium after it is dissolved in cooled rice water. The bacteria needs the carbohydrate base to react. The method was being implemented only in 1993. Every alternate year the land is ploughed.

The team was shown the plantation has trenches for drainage. The tree density is at 200 trees per ha *Teprocia purpurea* is found growing in small clumps. The ploughing activity, carried out every alternate year, perhaps uproots them so that they cease to spread. No undergrowth was observed except for occasional cacti and *euphorbia spp*. Leaf litter is very low. The plantations do not look healthy; the trees being thin and straggly. Grazing is permitted, but there is little to graze. No fences are put up. There is negligible pilferage and hardly any fire incidents. The productivity of the plantations was nowhere near the estimates. It is a wonder that so much wasteland has been covered with some tree cover but is it worthwhile?

At this point it is important to remember that FAO had recommended eucalyptus as an alternative to bamboo at two points of time. In 1959 the FAO/ESCAFE consultants recommended 50,000 acres were still available at Nilgiris and 73,000 acres at Kodaikanal for plantations of *E. globulus* and *A. melaoxylon* but no mention was made of *E. tereticornis. E. globulus* is a high elevation species.

Even earlier, when the FAO specialist Von Monroy had noted the need to replace bamboo by eucalyptus plantations, no species was specified. This was translated into action by the MP Forest Research Institutes (see Chapter 3), which started massive plantations of *E. tereticornis* in that state. In 1959, the Planning Commission gave a mandate for clear felling for 1 per cent of the most productive forest and replacing it with eucalyptus plantations (third five year plan) and in the fifth plan the plantation area was raised to 2 per cent of clear felled area in Forests (fifth five-year plan). The untried *E. tereticornis* plantations had failed to respond.

FAO started its assessment of the 1955 hardwood plantations it started in Latin America, Asia and Africa. The results were out by 1970, but were published only in 1980, in two volumes. vol. I (1980: 237) prepared by B. Kurkland in 1975, rejects *E. tereticornis* as it exhibits inferior growth and low resistance to drought. But vol. II (1980: 211) prepared by a Finnish team claims that *E. tereticornis* has been successfully raised in India. It qualifies that it can acclimatize at low zones, vol. II also adds that sulphate pulping requires relatively low alkali charges and that its pulp yield is low compared to average hardwood. Was TAFCORN confused by this contradictory finding and chose to go only by the findings reported in vol. II of FAO's *Forestry Paper*, no. 19, 1980? The earnestness of TAFCORN official is unmatched by the Forest Development Corporation of other states including Karnataka which confesses that eucalyptus plantation is a folly. The contradictory reporting in the two volumes of FAO's *Forestry Paper*, no. 19 became clear only in October 1995, when Kurkland himself explained in a very matter of fact way that the data on *E. tereticornis* in vol. II was mistakenly based on the findings in Latin America. The issue on *E. tereticornis* he added, was complicated due to numerous provenance of that species (personal communication, October 1995).

The GOI no doubt worried over the failure of eucalyptus to take off, especially since it was toying with the idea of starting its own units to produce not only foreign exchange saving newsprint but also cultural paper which the recalcitrant private sector was reluctant to produce (see Chapters 2, 3, 4 and 5). It commissioned FAO in 1970 to investigate this at a huge cost of Rs. 31.39 million. In its final report, FAO (1980: 29) states *E. tereticornis* which is widely planted has only limited use as pulpwood, and its coppice crop is progressively poorer in quality. It approved both *E. globulus* as well as *E. grandis* as worthy crops, though short-fibred (p. 25). FAO suggested *Gamelina aroborea* and *Albizzia spp.* and *pinus carabiacea* (p. 27) as alternatives. FAO's bureaucratic indifference to the destruction of a Third World's Forest ecosystem is not justified, but understandable as being impersonal, even if it took on the role of an advisor. Whereas that of the Government of India which failed to communicate these vital findings to the foresters of FDC is certainly not justifiable. GOI has failed to trust its foresters and prefers the opinion of FAO, an UN organization. It is a colossal drain on public exchequer—see Table 8.3.

Hopefully, the improved methodologies will, on the contrary, increase productivity, but with some eco-friendly species, so that the industry which has been repeatedly demanding land to raise their own plantations can be satisfied. TAFCORN officials say they do not see any adverse effect of eucalyptus.

But the fact of the matter remains that these plantations are being promoted on land which is already lateritic. These species do not in any way enrich the soil they grow on. It is beginning to show. After three rotations the yields were much below the expected levels. Will replanting on badly damaged soil yield any good? The idea of having a tree cover on degraded land must be to harvest the yield; also, the soil must rejuvenate itself, to be able to support more trees. By raising such exotics which are no longer controversial on already degraded land is the fertility of the soil being regenerated in any way?

Putting up a tree cover on degraded land is a sound idea but not for long. If we are thinking in terms of long-term sustenance then we have to give back to the system. We can enrich the soil with inorganic fertilizers but for how long, before there is any build-up?

Since TAFCORN operates outside the forest area now only in arid regions why does it not try cultivating agaves?

Agaves are seen growing well all along the railway tracts in dry areas and presumably grow well almost anywhere. Central American varieties have been imported in African countries to be planted in their arid region and they have done well (Lock 1962). They were imported for sisal based industries, which includes bituminised paper and fibre reinforced boards (Lock: 51, 284).

SECTION 3 (A)

Political Economy of Pulp and Paper Industry in Tamil Nadu

The origin and growth of pulp and paper industry in Tamil Nadu is different from that in other states. The reasons for this are historical. The Tamil-speaking area of the old composite Madras Presidency was relatively more developed than other linguistic areas within that state. As a result, at the time of transfer of power, Tamil Nadu had several well-established industries like locomotive, automobile, cement, confectionery of world repute, textiles and its ancillaries, such as tamarind starch for sizing, etc., to mention only a few. Besides it occupied a high rank amongst India's 27 states in education and literacy. It also enjoyed a reputation as the most intellectual state in the country. However, the state could not boast of a pulp and paper industry, a coveted one, that used to be the British preserve and which after the transfer of power, was elevated to the level of a core industry.

Investment for setting up a paper industry is enormous. The Natukotaichettiars, the wealthiest group, had resources but were mainly interested in finance and banking. In fact, in 1946 M. Ct. Chidambaram Chettiar set up Travancore Rayons in Travancore which was very attractive, but not in Tamil Nadu. By and large, industrial capital in India was largely invested by Marwaris, who had forged political links with the central government. The self-conscious south did not care to let a northerner invest in its state.

It is not very clear whether FAO which had entered India with a mandate to find tropical pulpwood and set up PPI units, was called in by the Madras Government for consultation or whether FAO had volunteered to help.

Its consultants suggested two sites—one near the Mettur dam. The rationale was—ample hydel power; easy disposal of effluents as the flow of river Cauvery is 30,000 cu.ft and in the driest season, 2,000-4,000 cu.ft; Mettur Chemical Co., produces caustic soda and chlorine; the railhead is Salem junction, with close proximity to bamboo sources—north Arcot, Nilgiris and south Coimbatore; blue gum and wattle from the Nilgiris and reeds received from Tirunelveli (FAO, 1962: vol. II, 35). The other suggested site was Nellikuppam which is connected by a medium gauge railway track, it has no river but water was said to be available 40 ft below surface, and effluents were to be disposed through gravitational pipelines into the sea. Both sites were totally unsuitable for a PPI unit.

To set up a water and power hungry PPI unit and one which was highly polluting in the command area of Mettur dam, the only one which irrigates vast areas in an uncertain and generally rain-starved state would have been a total disaster for the state. The canal waters of the Mettur dam are also used for domestic purposes in villages in downstream areas. What finally transpired is not clear but the Tamil Nadu government quietly rejected FAO's suggested sites. The first PPI unit the Seshasayee Paper Board (SPB), was installed at a location that was ideal from the industry's point of view. True, the river Cauvery would be polluted by the mill, which was located on its banks close to Erode; but a railway junction could bring in raw material from both the north and south tracks and transport back its finished goods; pollution was another matter and needed to be corrected.

With the linguistic state reorganization, the thickly-wooded areas of the Western Ghats were transferred to Kerala and Karnataka, and the best wooded Eastern Ghats to Andhra Pradesh, some being lost even as early as 1936 to Orissa. Nevertheless, the forest survey indicates that Madras state still had pockets of thickly-wooded areas in the Nilgiris and Anaimalais on the Western Ghats, on the Palani Hills which stretch eastwards from the main range and on the disjointed Shevaroy and Pachamalai hills on the Eastern Ghats. Sathyamangalam and Javathri hills are also a rich source of bamboos. To add to this, FAO suggested blue gum and wattle that the British colonizers had introduced as early as 1832 to provide firewood for its troops. They grew easily in the Nilgiris. Casuarina, another late discovery as a pulpable species, had also been introduced along the river banks, the sea coast and wherever else possible in the second half of the nineteenth century as fuel for locomotives.

Besides plantations, FAO also suggested a new alternative material, viz., bagasse—with details fully worked out to increase the output of sugar cane in the region and get control over the spent out cane, the bagasse. This will be discussed later. This is not to say bagasse is FAO's finding. Its use as a cellulosic material was known in India and was investigated by Bhargav way back in 1933.

Tamil Nadu had two other PPI units. It is believed that at the behest of Chief Minister M.G. Ramachandran the SPB set up the Tamil Nadu Newsprint Mill (TNPL) in the adjoining Trichy district, which also had the triple advantage of the perennial river Cauvery, the main well-laid track of the south Indian railway, and a vast area capable of being transformed into a sugar belt. SPB resources, even along with the equity offered by the state government, were too inadequate to set up a modern mega-newsprint plant. SPB negotiated with the World Bank and was able to get it more than interested in the project. The World Bank (WB) instead of offering a soft loan chose to invest in TNPL. Having entered, the WB used its clout to get 20 per cent eucalyptus from the Tamil Nadu government, apart from an assurance from the Andhra government for a steady supply of coal for its own boilers and from the railways for rakes to carry an uninterrupted flow of fuel. WB and FAO work in tandem. As early as 1959, FAO/ECAEFE consultants had noted that noted there were five sugar mills with an annual crushing capacity of 8,70,000 tonnes of sugar cane and four with a capacity of 4,50,000 tonnes. The total bagasse available would be 13,35,000 tonnes which would be equivalent to only 2,30,000 of air-dried bagasse. Since bagasse is used as fuel, by sugar mills, a supply of alternate fuel was critical. They suggested lignite which occurs near Neyveli be substituted as fuel for the sugar mills, 40 km around Neyveli.

First SBP, and the TNPL, changed the pattern of sugar cane

production. The old restrictions were thrown overboard; the sugar mills around the area were modernized with efficient boilers and supplied initially with oil and then with lignite to fire their boilers in exchange for the bagasse which was the traditional fuel used by sugar mills all over the country. TNPL, SPB's baby, cut its umbilical cord. For sometime it maintained its relationship with its parent industry as a consultative unit and then totally severed the relationship and forged ahead in every sense of the word, even in terms of the raw material that it acquired from the government. SPB is sore on this point. SPB production capacity has remained at 55,000 tpa since 1979, and its product line is also unchanged, whereas TNPL has plans to double its capacity by 1995. The civil engineering work is going on. It produces not only newsprint but writing and printing paper as well: the latter is a safer bet today with the changed economic policy.

True, investment for a newsprint mill is heavy, the gestation period long and the uncertainty of raw material keeps growing. But it still does not explain the reluctance of the local industrialists to enter the industry. Since 'north Indian Domination' and Marwari capital is abhorrent, the Tamil Nadu government chose an Italian multinational to align with it to produce viscose rayon.

Growth of viscose rayon is outside the scope of this study, as it does not in anyway add to the development of the country. But, what is of concern is the consequences of it drawing its raw material, whether from the natural forests or from plantations specially reared for its maintenance. The pulp for rayon weave is alleged to be finer than that of paper.

Several species of eucalyptus have been grown in India since 1832; the most popular is the *E. tereticornis; E. camaldulances* is perhaps a hardier species; it is said to be short-fibred but rejected by FAO (vol. II, 1980: 130) as unsuitable for Indian conditions, *E. globulus* that grows only on the high ranges is said to yield the best pulp. Both the paper and the newsprint industry allege that the Tamil Nadu Forest Department's priority is for Tamil Nadu Viscose also a joint sector venture and therefore the entire crop of *E. globulus* is given to the rayon industry. Our tabulated data presented later does not quite confirm this allegation. Rayon industrial units in Kerala, Karnataka and Andhra however, cannot claim such rights as none of them have such high ranges to cultivate *E. globulus*. Besides, the unit is also given priority to bamboo that is grown in Satyamangalam. We hold that rayon is not a socially useful product and should be dispensed with. SPB produces socially useful paper. In order to compete and survive, it aims to modernize, double its production and export its product.

TNPL produces socially useful newsprint and cultural paper. This unit, like SPB, aims at export, and, like it, plans to use the industrial effluents to irrigate sugar cane fields in its wide catchment area. The negative consequences of the chemicals—the effluents on the soil—the long run are discussed elsewhere. But what is important to remember is that a product that has potential for contributing to development should not have any negative features, or be so negative as to destroy the very base from which it draws its raw material. Besides, Tamil Nadu's tie-up is with the all-powerful World Bank which certainly does not have a reputation for preserving the ecosystem.

SECTION 3 (B)

Seshasayee Paper and Boards

Many states had built pulp and paper mills during the 1950s, but Tamil Nadu was without any, right up to 1960. At the express request of the Minister of Industry, R. Venkataraman, the Tamil Nadu government invited S. Viswanathan, a prominent labour lawyer in Madras to put up a pulp and paper mill. R. Venkataraman's appointment as Minister of Finance at the centre gave a fillip to the company. According to S. Vishwanathan the company was named after his friend and 'mentor' R. Seshasayee, a leading industrialist of Madras state who wanted to set up a paper mill but died in 1958, before his dream came true. The Seshasayee family invested in the mills. According to Baldwin, the Seshasayees are Tamil Brahmins of humble origin from Tiruchirapalli. The family set up was road transport operators in and around their home-town during the first world war. Soon after that one of the founding brothers, R. Seshasayee went to USA for training in electrical engineering. In the early 1920s he and his brother V. Seshasayee established themselves as electrical contractors. In 1924 they pioneered power generation and distribution in rural areas of Madras Presidency and founded the first three companies that were merged in 1940 to form the South Madras Electric Company. During the Second World War they took over Mettur Chemicals (that FAO had noticed), and ventured into prospecting for minerals but failed. The group acquired a minority interest in the Elphinstone Cotton Mills at Bombay. It specialized in the field of aluminium cables and from 1958 onwards, went into fertilizers, forest and chemical production in Travancore, and paper in Erode, Tamil Nadu (Baldwin 1959: 271). They ceased to be managing agents of the government company from 1968. R.K. Hazari lists Seshasayees as one of the six medium-sized among the 20 companies he examined.

Vishwanathan chose Erode as the site for the mill because of its proximity to the Nilgiris, a source of bamboo and hardwoods. In addition to this, the presence of the nearby Cauvery river proved to be a boon to fulfil the enormous perennial demand for water in paper production. Erode was also a railway junction, further easing out the problem of transportation of materials, raw as well as finished, for the mill. Thus, the location was perfectly suitable for the paper mill to prosper. The capital structure of SPB from 1960 onwards is as follows:

Capital Structure of SPB

Incorporation in 1960.

Issued 10,50,000 shares to promoters, etc., and collaborators and 10,00,000 shares to TIIC.

14,50,000 shares offered to public in September 1960.

Final call of Rs. 2.50 per share was made payable by July 1961.

17,50,000 bonus shares issued in proportion one : two (in 1976).

10,00,000 shares allotted on private placement basis in 1987-9.

10,00,000 shares allotted at par to financial institutions on 29 May 1989 in conversion of funded interest in 1989-90.

10 lakh equity shares of Rs. 10 each issued on private placement basis (of these 4 lakh shares issued to shareholders on preferential basis in the proportion of 20 per cent of their shareholding) (in 1991-2).

The Distribution of Equity Shareholdings of SPB (as on 30 July 1993) are:

Category	No. of shares held	% to total
I. Foreign holdings	602	0.04
II. Govt./Govt. sponsored Financial Institutions	4,28,645	25.51
III. Bodies Corporate (not covered under I & II)	8,57,303	51.03
IV. Directors & their relatives	0	0.00
V. Top 50 shareholders (not covered under I, II, III, IV)	92,224	5.49
VI. Others	3,01,226	17.93
TOTAL	16,80,000	100.00

Source: *Stock Exchange Official Directory Papers.*

The mill was built in collaboration with the leading US company, Parsons and Whittemore Inc., New York. The decision was taken in June 1960 and the first reel of paper was brought out in November 1962.

An article which appeared in the Financial Express (5 February 1979), gives another view of the mill's opening. According to it, SPB was sponsored as a joint venture by Seshasayee Brothers in 1960 in collaboration with the Madras Industrial Development Corporation, the US experts who certified its technical feasibility promising to deliver the turnkey plant and technical know-how and also subscription to its equity capital.

SPB's Operations

Seshasayee started commercial production with duplex board for packaging and unbleached varieties for carton making and production of laminates, using bamboo and softwood available in the vicinity of its factory, after conducting experiments on it. The company boasts that its entire production was sold out in the face of stiff competition in the year 1965-6 when the production figures stood at 20,250 tonnes. Encouraged by the results, the company made a small beginning in the export of printing quality paper (*Stock Exchange Official Directory*, 1988).

During 1966-7 and 1967-8 production increased marginally and then declined. The mill gives increased cost of inputs, power cuts and labour strikes as reasons for its poor performance. During the same years, the profits also came down mainly because the sale price of paper was inadequate to cover all these costs. The company tried to diversify into paper of lower grammage which fetches higher prices.

In 1968-9 the production curve of the mill took a upswing by 12 per cent and it exported 550 tonnes to Middle East countries. This year of good performance corresponds with the period (1968) when the price of paper was decontrolled and left completely open to the demand-supply forces.

This trend continued for the next five years (till March 1974) when the mill exported its products to Kuwait. There was a brief reversal during 1972-3 when production came down by 7 per cent. Most of the paper mills had taken to production of high value added speciality papers which led to acute shortage of cultural paper. But Seshasayee was not one of them. The shortage led to a rise in price. The government then fixed the price at Rs. 2,750 per tonne by an order in 1974.

Seshasayee, the unit under study, stated that production during 1975-6 rose although sales in terms of quantity declined. Other recalcitrant paper mills were forced to obey the production control order, resulting in a glut. Besides, owing to the general recession, there was an unprecedented decline in demand for paper. Consequently, allottees under the Paper (Control of Production) Order, 1974 failed to lift their goods. These two factors resulted in an accumulation of stocks and a sizeable downward trend in prices. The company's profit margins were therefore reduced. Seshasayee produced only cultural paper.

S. Narayanaswamy (1976), chairperson, stated in reference to this order, that 'There does not appear to be much logic in any industry being called upon to sell its output or any part thereof at below cost price'. It is evident from the consumption pattern of the previous year that the 'preferred sector' could do with substantially less paper than indicated by the present 30 per cent stipulation. Most of the time, the mills ignore one important point, that, apart from levy price and quantum levied (30 per cent), the mills are free to sell the residual 70 per cent of paper at prices regulated by the Law of Supply and Demand (chairperson's speech in the *Hindu*, 28 September 1976). It seems here too, that the big paper mills were at a disadvantage as there was a glut in the paper market due to a ban on exports (1975-6).

The year 1978-9 was claimed to be a notable one for the company by its chairperson, in that it was able to complete and commission its expansion programme embarked upon seven years earlier, in 1971, raising the production capacity from 35,000 to 50,000 tpa practically eight months ahead of schedule, at a cost of Rs. 17.5 crores as against the estimated outlay of Rs. 20 crores. The mill produced 47,841 tonnes of paper achieving 75 per cent of the rated capacity on the new machine during the first year of its operation. Unfortunately, the substantial increase in the cost of inputs, specially coal eroded its profitability by nearly 50 per cent (chairperson's speech, 1979).

The Paper (Control) Order, 1979 stipulated the price of cream wove/laid paper at Rs. 3,785 per tonne which was then being marketed at an ex-factory price of Rs. 5,300 per tonne. This evoked an adverse response from the paper mills (Rao 1989) to which Seshasayee was no exception. The SPB chairperson (1979) stated that this action of the government not only deprived the paper mills of the compensatory element needed to set off the loss on their supplies to government allottees, but also landed paper companies in deeper trouble by compelling them to incur losses in supply of cream wove varieties at an uneconomic price. The paper industry claims that it had no option but to go to court and ask for interim relief (chairperson's speech, 1979).

For the next six years (1979-80 to 1984-5), the *SEOD* reports record that SPB's operations were hampered by unprecedented levels of power cut, shortage of coal, workers' strikes and increase in the cost of chemicals, coal, power and railway freight. Is it a coincidence that all the large forest-based papers also complain of identical problems? It is interesting to see how in 1984-5, the mill diversified into Research and Consultancy Services.

The Directors seemed to be pleased to report in the year 1985-6 that the constraints which crippled the mill's profitability for the past four to five years were overcome to a great extent and it achieved a production of 55,202 tonnes, the highest ever in the history of the company (*AR*, 1985-6). Was it in anticipation of lifting of the Control Order?

However, the mill found it difficult to sell its output which led to an accumulation of higher stocks. Markets were reported to be in a sluggish condition. Apart from this, the company complained that the situation was further aggravated by small-scale paper mills which flooded the market with paper at lower prices. In the same report, the company mentioned that market resistance to its products was due to higher prices caused by ever-increasing cost of inputs, increase in excise duty and a compulsory short supply to government allottees at concessional rates. In this context, the management was engaged in taking energetic measures to make such varieties of paper which could absorb the hike in the cost of production. Also, the company constantly complained about the short supply of levy paper to government at concessional rates (*AR*, 1985-6). In 1985, the company came under the provisions of the Sick Industrial Companies (SP Provisions) Act 1985.

The experience of the government in regard to the short supply of white printing paper by the paper mills was in spite of periodic increases in its price from Rs. 2,750 per tonne in 1974 to Rs. 7,200 in 1986; also the forest-based paper mills were defaulting in supplying the requisite quantum of white printing paper. Therefore in January 1987 the production control order was withdrawn. The Seshasayee mill was happy and hoped that it would supply its product to the editorial sector. However, these expectations were belied when the government, simultaneously with the withdrawal of the production control order, entrusted Hindustan Paper Corporation (HPC) with the responsibility of meeting the entire requirements of the editorial sector with a subsidy support of Rs. 2,000 per tonne, over and above the price that was obtaining in the production control order at Rs. 7,200 per tonne (*SEOD*).

During 1988-9, production and sales declined further, resulting in losses. To take advantage of the increased demand for the cultural varieties of paper, the mill in 1988-9, gradually shifted its product mix to include more bleached cultural varieties in preference to duplex board and plain kraft which earlier accounted for over 60 per cent of total production. The mill found production of board and kraft paper unprofitable because of the rising costs of imported waste paper, an essential ingredient in production of these varieties (*AR*,1988-9).

Due to unprecedented power tripping and further increase in cost of inputs, the company incurred a loss of Rs. 153 lakh (*AR*, 1989-90). When fortunes of some of the stronger paper units in the country took a dramatic turn due to improved sales realization and removal of the paper control order, this company, like many others, did not show any great improvement in its working (*AR*, 1989-90).

The next year turned out to be a prosperous one for the SPB. It earned a net profit of Rs. 157 lakhs (as against a net loss of Rs. 153 lakh in the previous year). It turned the corner by increasing capacity utilization to 103 per cent, and with its added accent on increased production of low grammage bleached cultural varieties. Production for the year was 49,101 tonnes (at an average basis weight of 119 gsm). The foresightedness of low grammage proved itself in 1990-1 and the company geared up to produce good results in 1991-2. For the next three years the company put in determined efforts. Not only were costs controlled but a dramatic shift was made in its product mix so that realization per tonne of raw material improved substantially. The major steps taken by the company to make a comeback were:

1. Improved product mix with much higher realization per tonne on thinner varieties of paper.
2. Reduction in import dependence through Research and Development by shifting input raw material components from imported to indigenous ones.
3. Control of energy costs by installation of fuel efficient boiler; increasing the consumption of low-cost lignite and correspondingly halving of costly fuels like coals and furnace oil consumption. Thus the exposure of the company to price escalation in the oil sector could be reduced considerably.

4. The most significant part of the company's silent achievement was successful use of 100 per cent mixed hardwood and use of 85 per cent bagasse pulp for manufacture of cultural varieties of paper (*FE*, 23 September 1991).

The stock market's assumption that Seshasayee should be doing extraordinarily well in 1991-2, was rejected by Vishwanathan, the chairperson. According to him the paper market was not as buoyant as people made it out to be. Certain varieties like duplex and kraft paper were quite difficult to sell. The mill had to face competition from the small-scale sector. Because the latter did not have any excise liabilities it could afford to undercut Seshasayee's product in the market on the price front. 'If somebody is selling paper to you Rs. 2,000 a tonne cheaper than we are, why should you buy from us? This is a problem for us' explained the chairperson (Mudar Pathreya 1991). It is generally agreed that Seshasayee responded to the challenge by the small mills, by cleverly altering its product mix. The thicker variety of paper attracts mass production and as a result, Seshasayee has found it profitable to transit to the thinner grammage. This makes it possible for the company to produce more paper with the same amount of raw material, leading eventually to higher realization. It manufactures only very limited quantities of the low paying kraft and duplex varieties (Pathreya 1991).

For 1991-2 and 1992-3, the company achieved a high level of performance with an increase in capacity utilization from 125 to 133 per cent and an increase in net profit from Rs. 462 lakh to Rs. 628 lakhs.

The paper industry was being threatened by recession again. Though it increased production from 46,205 to 51,183 tonnes, sales declined marginally from 47,418 to 46,583 tonnes due to a sluggish market. However, the stress on production of low grammage (95 gsm) presently, and efforts to reduce cost of production still continue (*AR*, 1992-3). The production and sales Figures of SPB from 1962-3 onwards are represented in Figure 8.2.

The product mix of SPB has undergone a radical change. Traditionally, Seshasayee Paper and Board was a major supplier of duplex board and unbleached varieties, for carton making and production of laminates. The bleached varieties constituted only 38-43 per cent of the production, while presently the spacing production of fine papers accounts for nearly 80 per cent.

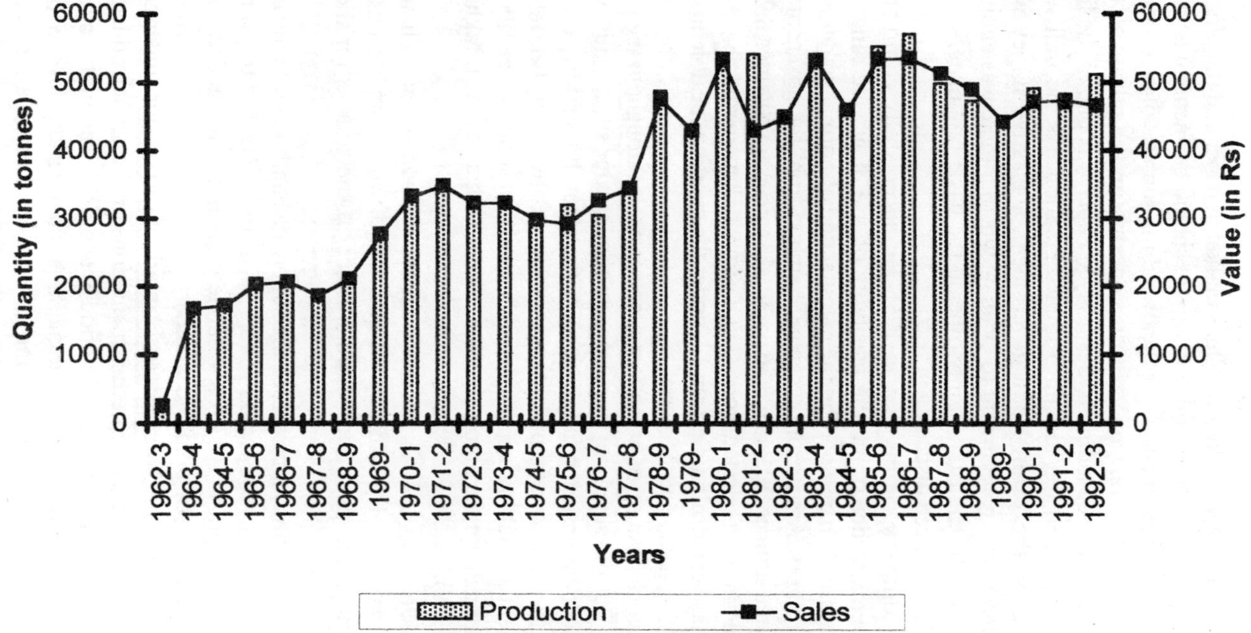

Source: Stock Exchange Official Directory

Figure 8.2: Production by and Sales of SPB (1962-3 to 1992-3)

Raw Material

Influenced by the growth of a seller's market in paper and the official thinking that the consumption of bagasse should be encouraged as a source of raw material to cover the dwindling supplies of traditional forest produce, several projects were conceived in the private sector at the end of the second plan. Seshasayee Paper was one of the few which transmitted the idea into implementation. As the mill was promoted by the Tamil Nadu government, a massive survey was conducted by the government to evaluate whether there was enough bamboo for the industry, bamboo being the basic raw material till 1978. This has been discussed in Section 2.

The company claims that in 1966-7 it successfully completed experiments to determine the feasibility of use of a wide range of locally available hardwoods in the manufacture of normal grades and types of paper. Apart from receiving bamboo from the government and private sources (30,946 + 8,367 = 39,313 tonnes), it obtained 8,715 air-dried tonnes of selected wood (*SEOD*) from unknown sources— this aspect has been dealt in Section 1—under the question 'head-loads of wood being taken out of forests'.

The stock position of bamboo was comfortable throughout the year 1967-8. But the supply of bagasse from the Pugalur sugar mills was inadequate for ensuring full working of the plant. It therefore sought to raise the capacity of the plant to 24,000 tonnes by adding balancing equipment and taking steps to depend entirely on supply of bamboos by shutting down the highly uneconomic pulping unit fed on bagasse (Gem 1979).

For 1968-9 to 1971-2, the raw material position of the mill was satisfactory. Table 8.6 indicates the quantum of wood supplied by the government and purchased through other private sources for these years.

It further had its contract for the supply of bamboo and pulpable wood (in 1971-2), renewed till March 1978, and persuaded the Tamil Nadu government to set up a separate corporation for undertaking intensive planting of pulpable wood.

In 1976, the Tamil Nadu Forest Plantation Corporation undertook growing of eucalyptus and other plantations on a commercial basis. The year coincided with the period when SPB was planning its expansion to 55,000 tpa. In this context, the chairperson pleaded for a joint effort by the industry and the government through a long-term lease of the required lands in an appropriate forest area, to the industrial

TABLE 8.6: RAW MATERIAL 1979-80 TO 1992-3

Year	Type of Wood	Forest Dept.	Social Forestry	TAFCORN
1	2	3	4	5
1979-80	*E. Hybrid*	19,092	—	19,44
	Frenella rhomnbodea	674	—	—
	Bamboo	29,412	—	—
1980-1	*E. hybrid*	18,115	—	39,538
	E. grandis	4,706	—	—
	Wattle	10,471	—	—
	Frenella rhomnbodea	1,063	—	—
	Bamboo	46,882	—	—
1981-2	*E. hybrid*	28,930	—	63,264
	E. grandis	—	—	—
	Bamboo	20,560	—	—
1982-3	*E. hybrid*	30,040	—	90,805
	Bamboo	24,411	—	—
1983-4	*E. hybrid*	30,768	—	94,981
	Bamboo	19,115	—	—
1984-5	*E. hybrid*	27,008	—	54,563
	E. grandis	11,549	—	—
	Wattle	2,275	—	—
	Pinus	6	—	—
	Bamboo	12,422	—	—
1985-6	*E. hybrid*	24,611	—	27,865
	E. grandis	8,133	—	—
	Wattle	11,852	—	—
	Pinus	40	—	—
	Bamboo	10,970	—	—
1986-7	*E. hybrid*	28,902	—	13,055
	E. grandis	2,379	—	—
	Wattle	10,250	—	—
	Bamboo	12,430	—	—
1987-8	*E. hybrid*	8,300	—	26,130
	E. grandis	1,862	—	—
	Wattle	14,155	—	—
	Bamboo	7,146	—	—
1988-9	*E. hybrid*	20,254	—	34,033
	Wattle	11,214	—	—
	Bamboo	2,377	—	—
1989-90	*E. hybrid*	11,552	—	40,173
	E. grandis	4,627	—	—
	Wattle	11,909	—	—
	Bamboo	2,000	—	—
1990-1	*E. hybrid*	5,365	—	35,982
	Wattle	13,385	—	—

(Contd.)

TABLE 8.6: (Contd.)

1	2	3	4	5
	Bamboo	6,536	—	—
1991-2	*E. hybrid*	3,334	8,243	27,430
	Wattle	12,173	—	—
	Pinus	9,897	—	—
	Bamboo	10,509	—	354
1992-3	*E. hybrid*	6,117	3,409	30,251
	Wattle	5,310	—	—
	Pinus	5,073	—	—
	Bamboo	5,625	—	1,227

Source: Seshasayee Paper & Boards, 1993.

units to grow their own pulpable varieties of wood which would greatly help the industry entering a new phase (chairperson's speech, 27 September 1976). But this was not granted.

In 1977-8 the state government raised the seigneurage rate for eucalyptus hybrid by 33 per cent to Rs. 80 per tonne when the company had sought a cut from Rs. 60 to Rs. 40. The company could not escape from the national trend just because the Tamil Nadu Industrial Development Corporation (TIDCO) had an interest in it (Gem 1979).

In 1985-6, the rates of royalty were hiked to a high level of 50 per cent in case of wood; it was 233 per cent in case of bamboo, which, along with the increase in cost of inputs, affected the mills operations financially. The company claims that another redeeming feature of production operations of the mill during this year was the use of unconventional raw materials, viz., bagasse (26,153 tonnes) and waste paper, which earned a substantial gross excise duty to the extent of Rs. 538 per tonne. However, the use of unconventional raw materials proved costlier to the mill as it had to incur additional capital expenditure. As a result, the net gain was only Rs. 233 per tonne (*AR*, 1985-6). Raw material consumption pattern from 1985 onwards is represented in Figure 8.3.

The production of bagasse pulp was 37 tonne per day in 1986-7 which later increased to 57 tpd in 1988-9. It was cited by the mill as a factor contributing to the increase in variable costs (*AR*, 1988-9). Compulsions to use high volume of bagasse pulp to offset the gap in availability of bamboo and wood in the pulping system designed to use forest-based raw materials, resulted in higher chemical lossess (*AR*, 1986-7).

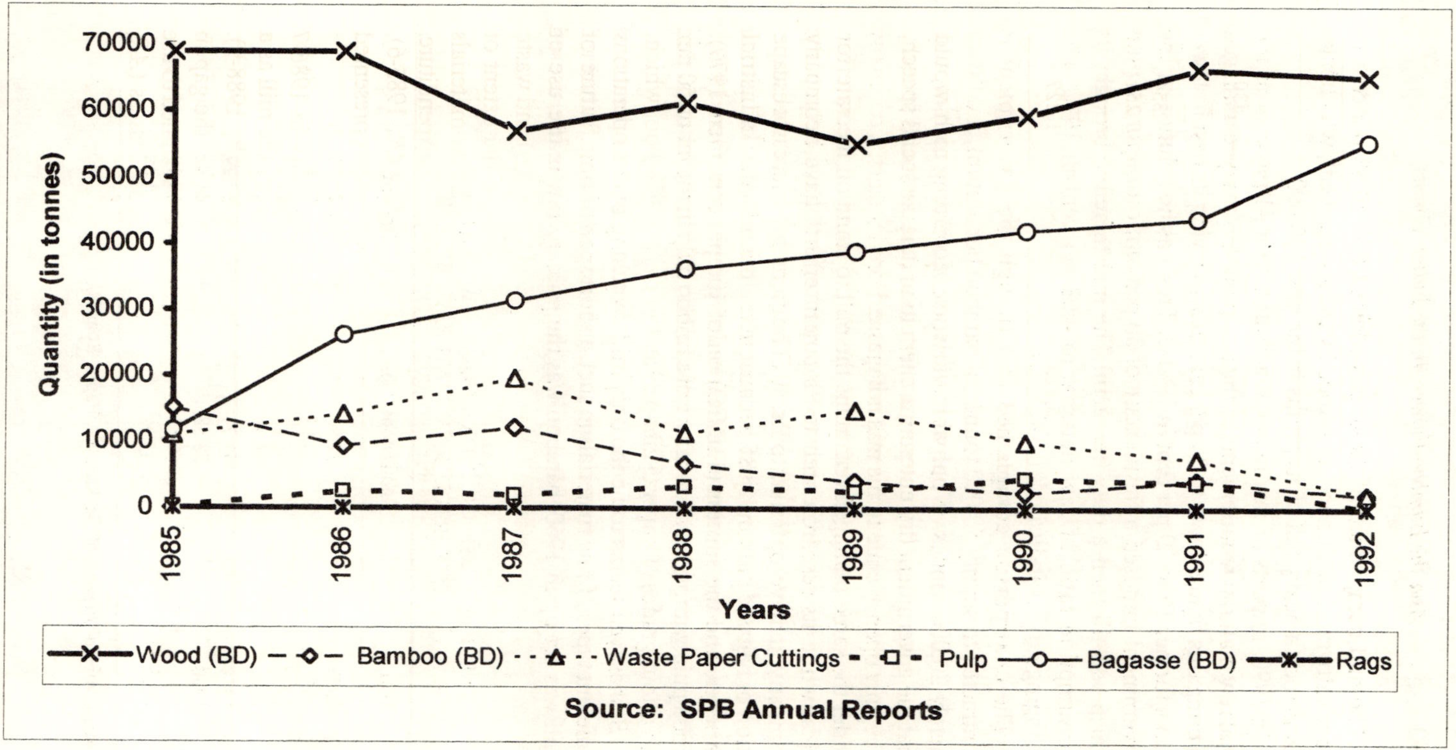

Figure 8.3: SPB's Consumption of Raw Material (1985 to 1992)

The mill complained of the unprecedented hike of 83 per cent in the royalty rates of bamboo and wood by the government, which hike it said took a heavy toll of its profits in the year 1989-90.

In 1990, imports of raw material accounted for 41 per cent of the total raw material requirements of the company. Seshasayee managed to bring this figure down to 30 per cent. It was trying its best to lower it to a further 15 to 20 per cent in 1991-2. This was no solution as the government imposed a fresh excise of 30 per cent on imported pulp which was till then a duty-free item. The mill therefore brought its consumption ratio of raw material to 4.08 per cent in 1992. The changing ratio is indicated in Table 8.7.

The chairperson complained that though the government is contracted to supply 50,000 tonnes of bamboo per annum, it delivers merely 20,000 tonnes. What was still worse, according to him, was that the government did not permit them to undertake social forestry. The only open avenue to the mill is bagasse. Ponni Sugars, a company under the same management, helps the mill to stand on its own feet. Meanwhile its research and development efforts have helped the company achieve optimum process conditions for successful utilization of bagasse pulp up to 85 per cent in the manufacture of cultural varieties, and the successful utilization of 100 per cent mixed hardwood pulp without any admixture of bamboo pulp in the manufacture of various grades of paper (Sinha 1991).

Seshasayee seems to be the only mill switching over to the use of different types of raw materials in such a short span of time. When the mill was set up in 1960, bamboo was the primary raw material used

TABLE 8.7: USE OF INDIGENOUS RAW MATERIAL AND IMPORTED RAW MATERIAL BY SPB

Years	Indigenous (%)	Imported (%)
1985	72.96	27.04
1986	67.99	32.01
1987	58.95	41.05
1988	63.70	36.30
1989	62.71	37.29
1990	58.58	41.42
1991	69.81	30.19
1992	95.92	4.08

Source: *Annual Reports* of Seshasayee Paper Boards.

and the mill machinery was designed to handle bamboo. During the second decade, the machines were forced to use tropical hardwoods to a high volume, which was again modified during the third decade to use substantial quantities of bagasse. It is not clear why this forward looking mill is not turning to its R and D to work on agave as well as on banana pith-based paper. The former has been used in Kenya (B. Kurkland, FAO, personal communication, October 1995). Tamil Nadu has a long history of using banana pith fibre to weave attractive shimmery silk-like fabrics. Both would yield strong fibre for paper boards if not paper.

Affluence from Effluents

Seshasayee Paper Mills credit themselves as being the first in India to introduce and implement the idea of manufacture of paper from bagasse. When they started out in 1962 their raw materials were bamboo and bagasse. At that time they got the bagasse from Rugalur Sugar Mills and in return, supplied oil to the sugar mills for their boilers.

By the late sixties there was an oil crisis and the price of oil rose steeply. The paper mill could not supply the oil. The agreement with the sugar mill was terminated and bagasse was replaced with eucalyptus wood.

The industry was not content with the raw material supply. The amount received was not enough (details in section on raw material). By the late 1970s alternatives to bamboo had to be sought. The answers were not so easily available since (a) As early as 1977-8 the Government of Tamil Nadu had put a ban on felling in natural areas. There were stringent orders which made raw material from natural areas inaccessible; (b) The Land Ceiling Act saw to it that lands were not available for the industry to raise their own plantations; (c) Schemes like farm forestry were not feasible in the nearby areas since the land holdings were small and it was not economical for the local farmer to wait for a period of eight to ten years for a tree crop.

The alternative was to go back to bagasse in a big way. The chairperson of SPB, Viswanathan, being on good terms with the state government and the Chief Minister, M.G. Ramachandran, was able to get the go-ahead to set up a newsprint unit based totally on bagasse. This saw the birth of TNPL in 1982. Along with this, ground work was on at SPB to supplement their raw material with bagasse. In 1984 SPB promoted Ponni Sugars, an associate company, located just next to it. Up to this time sugar mills came in to the cooperative sector. For a period of two years this law was eased, and SPB was therefore able to set up Ponni Sugars (*Company Sources*).

Ponni had some command area where it developed sugar cane. But the yield was not enough. Though the deep Cauvery runs through the area, there is not enough gravity flow and lift irrigation had to be taken up. Sugar cane requires a lot of water and these areas around were drylands and fallow lands, not suitable for sugar cane.

In 1984, SPB, Ponni Sugar and the local farmers entered upon a tripartite agreement (represented schematically)

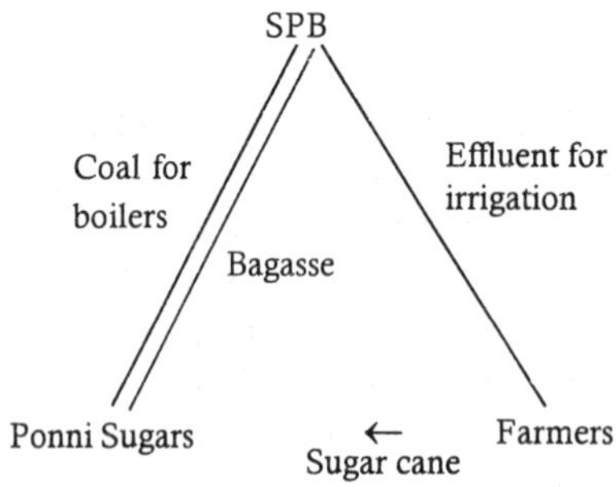

Diagram 8.1

The agreement saw to it that the farmer was happy, being able to raise a cash crop on his land which was till then fallow; the sugar mill got enough sugarcane and the paper mill got the bagasse.

The farmers paid Rs. 1,000 per year to SPB towards lift irrigation costs. They formed a 'Lift Irrigation Society' by which the four villages get irrigation facilities. Diagram 8.1 shows the scheme and the command area. This scheme resulted in an additional 50,000 tonnes of sugar cane, and also provided water to raise various crops (Table 8.8).

Over the years the farmers began to use the effluents, quietly, at first with the overspill, and then regularly as an irrigation scheme.

While the effluent is used for irrigation, it is necessary to consider what this will do to the soil in the future. Before launching the scheme, the mill approached Tamil Nadu Agricultural University (TNAU) and a scientist, a professor of macrobiology, who was appointed to monitor the areas regularly.

TABLE 8.8: CONSUMPTION & PATTERN OF WATER BY DIFFERENT CROPS

Crops	Area in acres	Quantity of water consumed cu.ml/sec.
Cane	1125.00	0.33
Paddy	187.50	0.06
Groundnut	187.50	0.03
Cane	33.98	0.08

The effluent used is subjected merely to a primary treatment. When it reaches the field after the primary treatment it still carries sulphates, sulphides and sodium compounds.

It would be worthwhile to highlight the findings of the Impact Assessment study done by TNAU. The aim was to assess the usefulness of the treated effluents for irrigating different crops, and the changes in the soil microflora, and macroflora, in relation to the physico-chemical properties of the effluents.

The effluent samples were collected from various discharge points and at various time-intervals (e.g. areas irrigated for 15 years, 3 years, 2 years, 1 year, etc.). The effluent had a pH of 8.7 (alkaline) in which the electrical conductivity was high. The primary treated effluent was low on plant nutrient. BOD was at permissible limits. Sodium content was high. The effluent was found to be 'good' in keeping with the standards required of effluents for irrigation.

The major findings of the study are briefly outlined below:

1. *Effect on Soil*: There was an increase in pH, EC and exchangeable sodium content of the soil and this increase was directly proportional to time. There was no significant increase in N, P & K contents of the soil (N, P, K—essential plant nutrients). While water-holding capacity and volume expansion on wetting were high in effluent-irrigated soils when compared to soil without effluent irrigation, the total porosity decreased in the former soil type. The fertility status of the soil was raised from low to medium. Effluent irrigation resulted in an increase in rhizosphere microflora of sugarcane and in the activity of soil enzymes.

2. *Effect of Treated and Untreated Effluent and Well Water*: A slight positive effect on germination, nutrient content, uptake, and chlorophyll centre were registered in plants irrigated for up to 15 years with treated effluents.

3. *Effects of the Flow of Effluents on Adjacent Soils*: This is as a result of seepage into neighbouring lands. There was an increase in sodium, nitrogen, organic carbon, calcium carbonate, pH, sodium and potassium per cent which was also indirectly related to the distance.

4. There was a reduction in the sugar cane yield in plots irrigated with the effluent for a period of 15 years and this could be attributed to the accumulation of salts in the soil.

In conclusion, the scientist who was consulted, says that there was a increase in soil pH, EC and sodium content which is directly proportional to the period under effluent irrigation. That there certainly was a boost in the micro-organism and enzyme content of the soil cannot be overlooked. But the fact of the matter remains that there was an essential build-up of sodium salts and calcium salts which were emanating through the surface soil and entering the subsoil. One can only wonder what the eventual effect of such a build-up would be.

For the present the effluent may boost the yield, but the effluent is not a mixture of plant nutrients. It is a mixture of what the industry cannot use and this need not be what the land can take in. The elements may be taken up but like all nutrient uptake, there is a prescribed dosage beyond which the element just accumulates.

Our discussions with the scientist who worked on the project threw open yet another aspect of the impact of the scheme. The area now under irrigation was primarily a dry area, irrigation by the river was not possible because of higher elevation where dryland crops like groundnut were raised. The land was not very productive. The lands at lower elevations were used for raising paddy. The small farmers were employed in the textile mills in and around the area. Prof. Dr. Obliswami recalled that when he went to the area 15 years ago, it was essentially a Gounder society in which the women controlled the land, worked on their fields and, to use his words, 'were dominant over the men'. They raised essential crops which met their subsistence needs from the land, while the men either worked on the lands or at the looms.

Now with enough water to irrigate their lands, the Gounders can raise cash crops. The land has begun to yield money and not just food as before. In such circumstance the male has assumed a dominant role since he controls the transactions with the mills. The women have become housewives, confined to their homes and hearths.

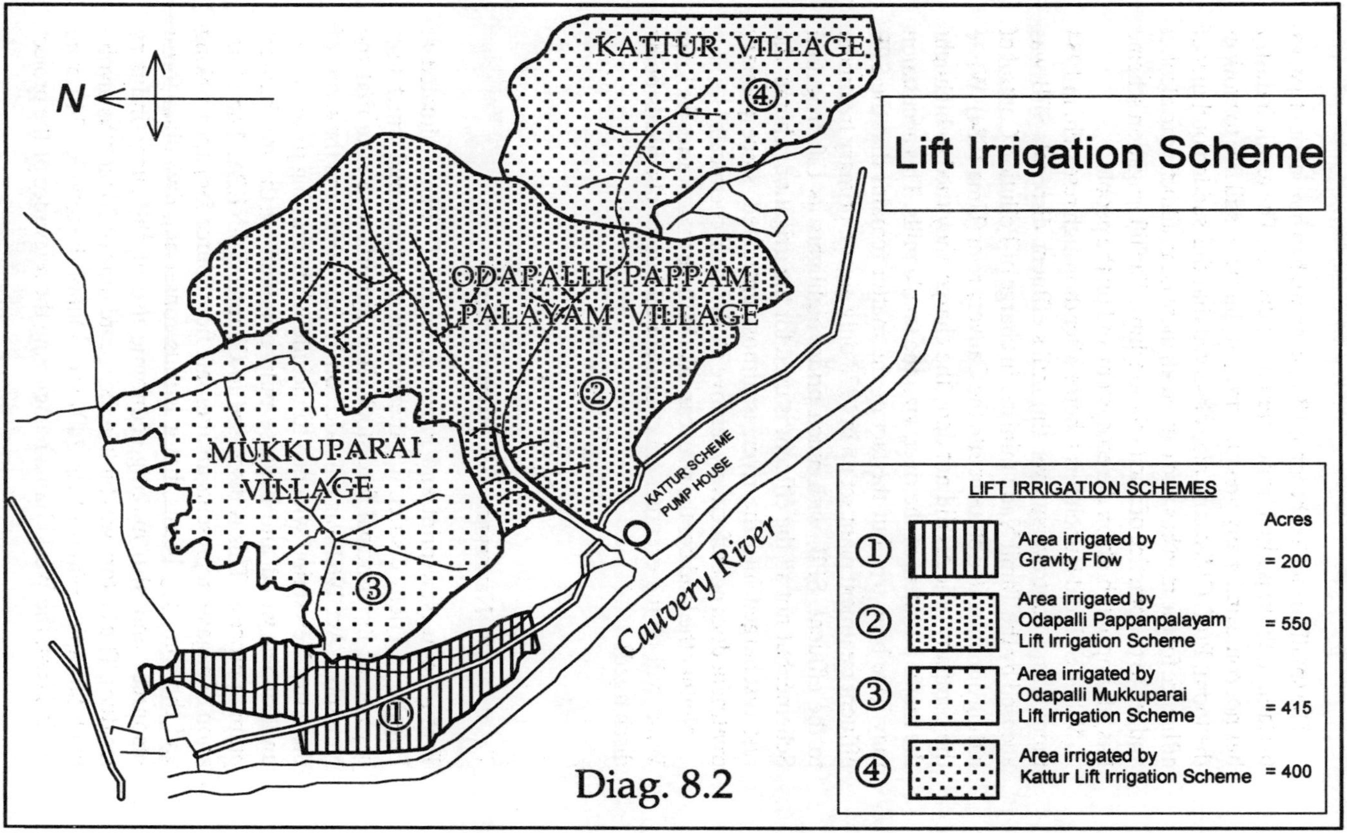
Lift Irrigation Scheme
N
KATTUR VILLAGE
ODAPALLI PAPPAM PALAYAM VILLAGE
MUKKUPARAI VILLAGE
KATTUR SCHEME PUMP HOUSE
Cauvery River
LIFT IRRIGATION SCHEMES
Acres
① Area irrigated by Gravity Flow = 200
② Area irrigated by Odapalli Pappanpalayam Lift Irrigation Scheme = 550
③ Area irrigated by Odapalli Mukkuparai Lift Irrigation Scheme = 415
④ Area irrigated by Kattur Lift Irrigation Scheme = 400

Diag. 8.2

The small farmer has become affluent enough to hire labour to work his land. Labour is hired from Thanjavoor district. The textile industry has not come under any pressure since the farmer is still free to work in the loom. Prof. Obliswami also observed that this sudden outburst of affluence for the men has also caused an increase in addiction to alcohol and led to the practice of illicit relationships. The lift irrigation scheme has left imprints both on the ecosystem and on the people.

While SPB did launch this scheme after thorough research, in 1984 there was a controversy over the mill's effluent disposal. SPB was accused by the Tamil Nadu minister in charge of Pollution Control of letting out untreated effluent into the Cauvery river (Uma 1984). While the company officials did not deny the charge, they tried to highlight the major polluters of the river, viz., the textile mills. This accusation may have been true but the fact of the matter remains that the SPB effluent treatment plant set up in 1978 only gives a primary treatment to the effluent. SPB, with much pride, highlights its Lift Irrigation Scheme; but not all the effluent is used for the irrigation scheme; the rest discharged into the river, is almost untreated, since primary treatment does not effectively remove all the toxic chemicals.

Neither the land nor the river can take in these toxic wastes for long. What will be the eventual effect of the build-up of chemicals being discharged?

SECTION 3 (C)

Tamil Nadu Newsprint and Papers Limited

Tamil Nadu Newsprint and Papers Limited (TNPL) is situated at Pugalur in Trichy district. It is the premier newsprint mill which uses bagasse as its principal feed-stock. It is important to note that the Newsprint Act 1962 was amended on 31 July 1992, and the scope of the term 'Newsprint' widened to include white printing paper, using unbleached wood and bagasse pulp obtained by the mechanical pulping process. This is discussed in Chapter 3—NEPA. The utilization of bagasse represents an alternate fibre source for paper making and the company boasts that it is thus conserving dwindling forest resources and also making economic use of the sugar industry residues. The factory was built in the record time of 36 months involving total capital outlay of Rs. 240 crore. Out of the total Rs. 240 crore capital cost, the World Bank, recognizing the relevance of the project to the Indian Pulp and Paper Industry and other sugar cane producing countries, gave the major portion of the project cost, i.e. loan assistance of Rs. 100 crore through IDBI. Another Rs. 55 crore came as loans from various financial institutions, and the remainder (Rs. 80 crores) was made up of equity held by the Tamil Nadu government.

Located on the banks of the river Cauvery, this plant has the capacity to manufacture 83,000 tonnes of newsprint and 17,000 tonnes of writing and printing paper per annum. It was originally designed to produce 300 tonnes of newsprint per day at 85 per cent efficiency or 240 tonnes of fine paper at 80 per cent efficiency (*The Times of India*, 1985). TNPL was designed to produce newsprint from 'mechanical bagasse pulp' on a commercial scale for the first time in the world. Bagasse and various varieties of hardwood pulp were also to be used in suitable proportions both in the manufacture of newsprint as also writing and printing paper. For production of these, TNPL has a single dual-purpose machine supplied by Beloit Walmsley (UK). Constant adjustment on this single machine to produce newsprint as well as writing and printing paper causes a tremendous amount of waste. This waste paper is recycled, but recycling implies another dose of chemical inputs, particularly power, which is a scarce commodity in Tamil Nadu.

The credit for the birth of TNPL goes to Seshasayee Paper Boards. TNPL is the brain child of S. Viswanathan, Managing Director of SPB. TNPL, however, gives SPB credit only as consultant to the project up to 1988.

In TNPL's capital structure, the Tamil Nadu government holds 49 per cent of the equity, the IDBI (in fact World Bank) another 49 per cent and the remaming two per cent is held by various state government undertakings (Pani 1990).

Operations of TNPL

Embarking on the project's spadework in September 1981 under a cloud of political and public controversy over such aspects as the inadequacy of the state-of-the-art technology and bagasse procurement hurdles, TNPL prides itself on having more or less kept its 'tryst with destiny', by commencing commercial production of newsprint in September 1984 on schedule (*FE*, 23 May 1986).

Significantly, it was only in March 1984, nearly two and a half years after work on the TNPL project commenced, that a Beloit

Seshasayee Paper and Boards (SPB) team is said to have successfully demonstrated, apparently on a pilot scale and of course to the satisfaction of the World Bank, the efficacy of producing newsprint from bagasse pulp, containing a significant proportion of 'mechanical bagasse pulp' as distinct from other varieties of pulp from the same species.

However, what is generally not known is that production started with imported and not indigenous pulp, a situation which continued till October 1985 and was maintained as a political secret (a senior FAO official, personal communication, October 1995). Also, what was imported was not bagasse pulp but only conventional wood pulp. Hardwood needed a chemical component to be added to classical thermo-mechanical technology. By October 1985, 'TNPL had fully acquired the capability to produce newsprint using 100 per cent bagasse pulp developed indigenously at what is reported as TNPL's own facilities (*FE*, 23 May 1986).

Against an annual installed capacity of one lakh tonnes of newsprint or 80,000 tonnes of printing and writing paper, TNPL opted from the beginning for a product mix of 50,000 tonnes of newsprint and 40,000 tonnes of writing and printing paper. It is not clear how it managed this. As stated earlier, permission for a newsprint mill to produce white printing and writing paper was passed only in April 1992. The World Bank having the clout to do so had the 1962 Newsprint Act amended. Reviewed in the light of such a product mix, TNPL is said to have attained 73 per cent capacity utilization in 1985-6.

In July 1985, TNPL produced 7,000 tonnes of newsprint—300 tonnes per day (tpd)—achieving 75 per cent capacity utilization, using 60 per cent imported and 40 per cent indigenous pulp. Production of writing and printing paper was said to be expected to commence from the second half of 1985 (*ET*, 2 August 1985).

It was only in October 1985, that TNPL could put Beloit-SPB's know-how to productive use, the time-lag being said to be apparently traceable to the inevitable intricacies of a critical know-how absorption. In 1986 TNPL used a pulp mix in the following ratio for the production of newsprint and writing and printing paper:

	Newsprint	Writing & Printing Paper
Mechanical Bagasse	50 %	nil
Chemical Bagasse	35 %	75 %
Hardwood	15 %	25 %

Of these pulp varieties, the use of mechanical bagasse pulp for the production of newsprint was said to be the most critical aspect in view of its uniqueness in the world newsprint scene (*FE*, 23 May 1986). The chemical bagasse pulp used in the production of writing and printing paper was produced by the sulphate process (TNPL Executive, personal communication, August 1993). Up to September 1987, the company had sustained a loss of Rs. 40.25 crore during the first three years of operation.

TNPL produced 76,597 mt and 50,076 mt of newsprint and writing and printing paper respectively, during the period of 18 months from 1987-9, achieving 94 per cent capacity utilization. According to the company's *Annual Report*, it sustained a net loss of Rs. 9.70 crore during the same period. As TNPL used bagasse, it was unable to achieve the required whiteness to match HNL's newsprint. But *The Times of India* assisted it in improving its product, and is producing pink newsprint exclusively for its *Economic Times* (*The Times of India*, Executive, personal communication, February 1994).

During 1989-90 there was a marked improvement in TNPL's performance. Out of 93,966 mt of total production newsprint production, was to the tune of 45,005 mt and writing and printing paper production was 48,961 mt, thus achieving a capacity utilization of 104.4 per cent. On the profit front, TNPL made a net profit of Rs. 31.08 crore while the cumulative net loss as on 31 March 1990, came down to Rs. 18.97 crore. This, according to the company's *Annual Report* 1989-90, was due to the increase in its sales achieved by high quality of writing and printing paper (*AR*, 1989-90). Figure 8.4 shows production and sales of TNPL from 1987 onwards.

In 1990-1, TNPL's *Annual Report* showed an increase of Rs. 5 crore in its net profit, touching the figure of Rs. 36.44 crore (from Rs. 31.08 crore in 1989-90). However, compared to the previous year, production showed a marginal decline. The production was 90,300 mt (93,966 mt) consisting of 39,064 mt of newsprint and 51,236 mt of writing and printing paper. The plant achieved 100.3 per cent capacity utilization due to increased production of cultural varieties of paper. The company proudly stated that the consumers acclaimed the quality standards of TNPL's products, consequently the market for writing and printing paper continued to be good. Despite high production levels, TNPL had managed to achieve 'zero stock' level of finished products at the end of the financial year (1989-90), for the second consecutive year. This does not necessarily imply that it is 'displaying

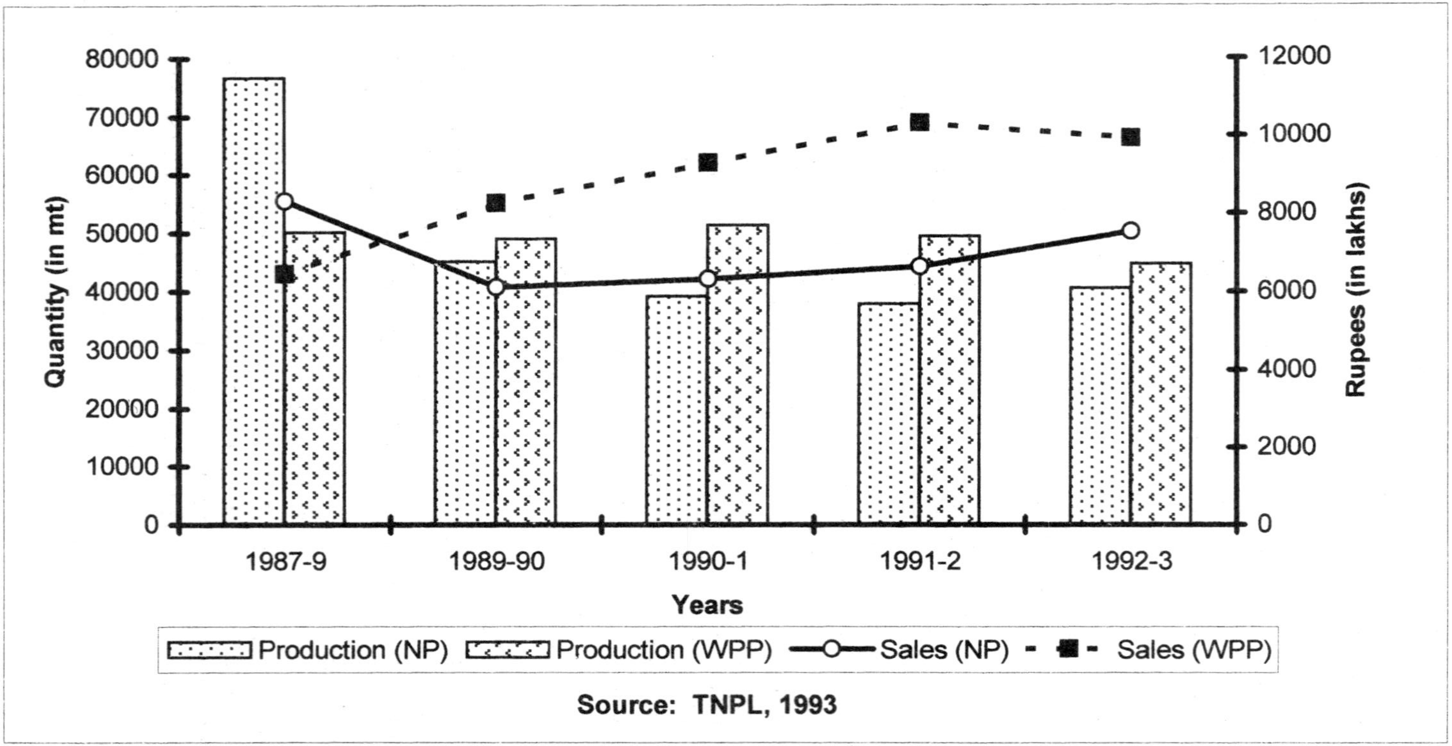

Figure 8.4: Production and Sales of TNPL (1987-9 to 1992-3)

marketing skill of a high order' as its *Annual Report* for 1990-1 claims. The situation has to be viewed in the context of the global newsprint scenario. The newsprint industry in the west had recovered, its price was rising, and it found no reason to undercut and dump its products in India. TNPL had no reason to feel complacent. From its inception in 1985 TNPL had produced 3.28 lakh tonnes of newsprint up to the end of March 1991. A higher yield of newsprint meant saving precious foreign exchange to the extent of Rs. 354 crore for the country at that point of time (*AR*, 1991 and *FE*, June 1991).

But in 1991-2 the production of TNPL declined to 87,242 mt (from previous production of 90,300 mt), consisting of 37,840 mt of newsprint and 49,402 mt of writing and printing paper, utilizing 97 per cent of capacity. TNPL in its annual report (1991-2) states that it responded positively to the change in the market trend for the lower GSM paper and that this resulted in a marginal drop in the physical weight without affecting profitability. TNPL maintained the record of 'zero stock' of finished goods for the third consecutive year.

In 1992-3 the plant achieved 97 per cent capacity utilization. The company was able to improve yields through the modification of mechanical pulping, where Thermo-Mechanical Pulping (TMP) refiners were arranged in a series with Chemi-Thermo-Mechanical Pulping (CTMP) refiners in order to cope with the addition of eucalyptus in the furnish. Thus, TMP an advanced technology which yields 95 per cent of the cellulosic material was completely eliminated, and the entire pulp for newsprint is manufactured by the Chemi-Thermo-Mechanical Pulping process and cellulosic yield from the raw material dropped to 85 to 90 per cent. JIC states that chemical pulp is used for writing and printing paper which is ecologically wasteful as cellulosic yield drops to 40 per cent. In 1992-3, total production was 87,242 mt consisting of 37,480 mt of newsprint and 49,402 mt of writing and printing paper. The on-stream efficiency of paper machines is indicative of the overall efficiency of the mill. At TNPL the average on-stream efficiency is 303 days, where 330 operational days are opted for calculating annual capacities. Out of these 303 total operational days at TNPL, 186 operational days have been allocated for the production of newsprint (*Journal of Industry and Commerce* [*JIC*]). The company proudly states that its products continued to enjoy a premium in the market but this is not confirmed by the consumers. TNPL also claimed that the good reputation of its quality products enabled the company to sell out the entire production achieving a 'zero stock', of finished goods for an unprecedented fourth consecutive year and this in 1992-3, which it described as a year of 'gloom and glut' (*AR*, 1993).

The situation changed drastically in 1993 with the onset of recessionary conditions in newsprint exporting countries. Foreign suppliers have reduced their prices from US $ 700 to 800 per mt during 1988-9 to $ 360 to 420 per mt in 1994; these are considerably lower than the cost of production of pulp in India. The reason for the drastic drop in newsprint price is discussed under HNL, Chapter 8.

TNPL's Product

TNPL was founded to produce newsprint with an in-built flexibility to also produce writing and printing paper. The corporate report on TNPL (Pani 1990) states that TNPL readjusted the product mix to take advantage of favourable market conditions in 1989 for writing and printing paper. In the writing paper segment a conscious effort was made in 1989-90, to enter into remunerative areas like computer stationery, lottery printing paper (LPs) and also photocopying paper, soap wrappers, etc. It also claims that the company, with its large and highly computerized machines, was technically better placed than most to take advantage of the boom in the paper market during the same year. Printers of computer stationery and soap wrappers, for example, preferred paper in reels rather than in sheet form. 'It was done to reduce wastage', and TNPL claimed that it is at a distinct advantage because its large machines provide reels with more paper than any other mill in the country, that its grammage was consistent and that the paper had fewer joints to prevent breakage. It is because of these reasons that TNPL claims that its products were in demand in the market, resulting in the 'zero stock' position of TNPL for four consecutive years from 1990 to 1993.

The company's entry into value-added papers has been an unqualified success, reports Dharani K. Pani (1990). Computer stationery, into which TNPL made an entry in 1986, has a market share of 80 per cent in the Bombay, Bangalore and Hyderabad markets. It has also been successful in marketing its lottery printing paper. TNPL has 50 per cent share of the 600 mt—Delhi lottery printing paper market, which is the biggest in the country.

In the area of newsprint, Chandresh Shah, a Bombay newsprint dealer says (see Pani 1990), that the quality of TNPL has improved considerably. Though its price is 10 to 20 per cent higher than the market rate, according to Pani (1990), TNPL's newsprint is preferred he says to newsprint produced by NEPA, Hindustan Newsprint Limited, Kerala or Mysore Newsprint, because of its high quality in terms of whiteness, bursting strength and grammage. This however, is not confirmed by our findings at our meeting with the representatives of newspaper industry.

Its latest breakthrough is said to be in the production of glazed newsprint introduced for the first time in the market by TNPL. The credit goes to the well-equipped, sophisticated, R and D wing of TNPL. Major buyers of TNPL newsprint are *The Times of India* and some regional papers (TNPL Executive, personal communication, Aug. 1993). It also has an agent in New Delhi who happens to be an indenting agent for imported newsprint and imported paper.

The list of different varieties of newsprint and paper produced by TNPL is given below:

I. Newsprint: Year of Introduction, November 1984
(a) Standard Newsprint (b) Pink Newsprint (c) Newsprint—High Bright (d) Newsprint—Non-Standard (e) Newsprint—High Bright, Non-Standard.
Average price: Rs.19,620 per tonne.

II. Writing and Printing: Year of Introduction, August 1985
(a) Cream Wove (b) Maplitho (c) Offset (d) Telephone Grade (e) Duplicating (f) Pink Wove (g) Lottery Printing (h) Super Lottery Printing (i) Special Printing (j) Yellow Wove (k) Cream Wove (Grey) (l) Cream Wove Natural Shade (m) Smooth Shade Maplitho (n) Maplitho (o) Offsite (p) Mirror Deluxe (q) Cream Wove Medium
Average Price: Sheets - Rs. 24,960, Reels - Rs. 24,155.

Source: TNPL.

Price of TNPL's Newsprint

The company states that an important factor to be taken into account while fixing the price of its newsprint, is that it is an import saving project. It was based on the principle of domestic resource cost (DRC) which meant that its end-product price could be 25 per cent more than the landed cost of the comparable imported product. This was necessary to establish TNPL's commercial success and also as an incentive for setting up projects of similar nature. 'Keeping this in mind', the union government exempted 100 per cent excise duty on paper made with more than 75 per cent bagasse in 1985. However, in the current union budget (1993-4), 10 per cent excise duty has been slapped on non-conventional raw material thereby forcing TNPL's product to compete with other mills.

TNPL's cost of production of newsprint is said to have worked out to Rs. 10,000 plus per tonne in 1986. The price, as fixed by the Bureau of Industrial Costs and Prices (BICP) and the Centre, was about Rs. 8,960 per tonne for all domestic newsprint. C. Ramachandran, TNPL's Managing Director (1986), pointed out that TNPL had asked (BICP) to consider fixing a separate price for the TNPL's newsprint in view of technological and other considerations, as also to consider a system of pooling imported and indigenous newsprint, as in the case of fertilizers, before fixing a common price. These steps would have stood TNPL in good stead, he argued, in view of its high capital cost which was at least Rs. 70 crore more than its nearest rival on the domestic scene (*FE*, 3 May 1986).

In 1990, *Business India's* Corporate Report on TNPL gives the viewpoint of R. Kuppu Rao, Deputy General Manager, that the price fixed for TNPL newsprint was inadequate to cover increases in the price of coal, railway freight, power and other inputs as well as government levies. Consequently TNPL's balance sheet showed consistent losses (Pani 1990).

TNPL's *Annual Report* for the year 1993 says due to the Government of India's policy of allowing duty-free import of newsprint on the basis of the 2:1 ratio and permitting small newspapers to import up to 200 tonnes of newsprint without insisting on domestic purchases has led to a crisis in the Newsprint Industry. This policy, it says, in the absence of proper monitoring mechanism, has led to the import of a large quantity of newsprint into the country during 1992-3. The Indian Newsprint Manufacturers Association has made representations to the government, making several constructive suggestions for taking care of the long-term interests of the newspaper industry as well as the newsprint manufacturing industry.

> We expect some policy changes soon. However, TNPL having some flexibility to manufacture printing and writing paper also, has decided to use this flexibility and steer clear of the newsprint crisis, which annoys the government. It plans to reduce its NP production to the minimum level and increasing the production of writing and printing paper. This strategy will help the company to maintain the profitability. (*AR*, 1993: 4)

These are 'words of wisdom'. The private corporate sector too, now with the liberalization policy, is planning to regress and produce only printing and writing paper as it cannot expect to compete with the speciality paper produced by large multinational paper corporations (see Chapter 4 for discussion on this issue).

Such paradoxical situations are evident only in the Indian economy where the government gives full support and encouragement to the industry at the initial stages, but once commercial production starts, throttles it ruthlessly by policy changes to such an extent, that the industry is forced to reduce the production of that item for which it was primarily set up. In the 1993-4 budget the government of India imposed a 10 per cent excise duty on non-conventional raw material thereby leaving TNPL's products to compete with imported newsprint dumped in India.

Contractual Agreement for Raw Material—Bagasse

TNPL is the country's first bagasse-based newsprint factory. The project was conceived in 1981 with an annual requirement of 4.20 lakh tonnes of bagasse on full production. Bagasse mainly constitutes fibre (48.5 per cent) and moisture (48 per cent). This brings us to the related question of procurement of bagasse. TNPL entered into separate agreements for bagasse procurement with 5 sugar mills within a radius of 5 to 140 km from its plant site at Pugalur in Trichy district. To cajole these sugar mills to part with bagasse which they otherwise used in their own boilers to produce process steam for their own requirement, TNPL set up 11 boilers designed for the use of both coal and lignite at a capital cost of Rs. 20 crore. The boilers were operated by TNPL's own staff. TNPL is given bagasse in proportion to the steam generated in the boilers set up in the sugar mills. The procurement of coal and/or lignite to run these boilers is left entirely to TNPL's care. On the whole, TNPL has to spend on an average, about Rs. 400 for obtaining a tonne of bagasse through this kind of barter of process steam for bagasse, but it includes the transportation charges and the investment on the bagasse-handling system at TNPL's own premises.

However, to meet any unforeseen contingencies, TNPL has chosen not to dismantle the existing bagasse-fuelled boilers in these five sugar mills.

TNPL's own project configuration now left it with an annual requirement of about 70,000 tonnes of eucalyptus, for procuring which TNPL entered into an agreement with the Tamil Nadu Forest Plantation

TABLE 8.9: PATTERN OF FOREST-BASED RAW MATERIAL PROCUREMENT (1984-5 TO 1992-3)

Year	Type of Wood	Forest Dept. (mt)	FDC (mt)
1984-5	*E. grandis*	21,320	—
1985-6	*E. hybrid*	—	39,674
1986-7	*E. hybrid*	—	52,754
	Wattle	11,250	—
1987-8	*E. hybrid*	—	32,500
	Wattle	15,000	—
1988-9	*E. hybrid*	—	43,089
	Wattle	14,355	—
	Pine	500	—
1989-90	*E. hybrid*	—	53,751
	E. grandis	4,030	—
	Wattle Wood	20,000	—
1990-1	*E. hybrid*	10,845	40,089
	Wattle	10,374	—
	E. grandis	13,952	—
	Pine	576	—
1991-2	*E. hybrid*	4,259	26,399
	E. grandis	6,842	—
	Wattle	16,353	—
1992-3	*E. hybrid*	13,466	32,250
	E. grandis	5,135	—
	Wattle	6,859	—

Source: TNPL, 1993.

Corporation. The procurement of raw materials from Tamil Nadu Forest Department and State Forest Development Corporation by TNPL is shown in Table 8.9.

If for some reason TNPL should use only eucalyptus giving up the bagasse route, C. Ramachandran, TNPL's Managing Director (1986), states that about 1.70 lakh acres might be needed to raise plantations to meet its needs. Such a large expanse of land is just not available for this purpose in Tamil Nadu. Asked if TNPL could itself go in for captive sugar cane cultivation and set up sugar mill(s), C. Ramachandran said TNPL was considering the promotion of sugar cane cultivation in a nearby area of about 1,500 to 2,000 acres. The yield when processed would give TNPL bagasse of the order of 1.20 lakh tonnes per annum. He did not go in to the specifics of a sugar mill in this context (*FE*, 26 May 1986).

TNPL is at an advantage over other newsprint mills since its major raw materials is bagasse the price of which has remained constant.

TABLE 8.10: CONSUMPTION OF RAW MATERIAL BY TNPL

(in tonnes)

Year	Wood	Bagasse	Imported Wood Pulp
1987-9 (18 months)	95,586	4,54,286	24,455
1989-90	68,945	3,57,584	12,903
1990-1	65,331	3,25,053	13,457
1991-2	64,789	3,08,061	11,998
1992-3	71,564	2,36,834	8,335

Source: *Annual Reports* of TNPL—1989-90, 1990-1, 1991-2, 1992-3.

Again, unlike other mills which have to transport raw materials from long distances, it is supplied by the State Forest Department with bamboos, waste and TAFCORN's eucalyptus. Even though bagasse is primarily an agricultural waste product, it is important to take its quality into account. It is said that the stones and sand in bagasse pose a problem at TNPL's mechanical bagasse-pulping line. Apart from this, bagasse is also believed to deteriorate through storage, causing a strong odour. This is very true. In the ultimate analysis, such a product results in some degree of dimming of the brightness of the end-product newsprint and other paper. Presently, bagasse pulp accounts for 85 per cent in the newsprint furnish and 75 per cent in writing and printing paper furnish, while hardwood (eucalyptus) and other tropical imported wood pulp account for the remaming 25 per cent. In 1993-4 the percentage of imported wood pulp was 34. The dust loss

TABLE 8.11: VALUE OF RAW MATERIALS USED BY TNPL

(qty. in mt) *(value: Rs. in lakhs)*

Year	Imported [Value]	Indigenous [Value]
1987-9 (18 months)	34.6 2% [1603.74]	65.38% [3028.34]
1989-90	28.7 8% [977.00]	71.22 % [2418.25]
1990-1	29.03 % [955.91]	70.97 % [2337.46]
1991-2	28.18 % [983.00]	71.82 % [2504.98]
1992-3	27.86 % [892.65]	72.14 % [2311.19]

Source: TNPL.

during the use of different cellulosic raw materials like wood, bagasse and eucalyptus as reported by TNPL is 4.5 per cent which is high; it should be possible to restrict it to about 2 per cent.

The actual consumption of different types of raw material since inception and their values are shown in Tables 8.10 and 8.11 respectively.

Apart from these procured raw materials TNPL has been buying eucalyptus wood from the open market. The quantity is shown in Table 8.12.

Coal: For its requirement the company has tied up with Coal India which supplies 1,90,000 tonnes and with Singareni Collieries in Andhra Pradesh which supplies 1,50,000 tonnes per year. In times of coal shortages, TNPL uses lignite from the nearby Neyveli Lignite Corporation.

Power: The use of bagasse for mechanical pulp is a pioneering effort by TNPL. However, the production of bagasse mechanical pulp involves high power consumption, while the yield of pulp is comparatively low (*Journal of Industries and Commerce*, May 1993: 9). The plant has two 25 mw turbo generators which meet two-thirds of its requirements, the balance being supplied by the Tamil Nadu Electricity

TABLE 8.12: OPEN MARKET PROCUREMENT OF RAW MATERIAL FROM INCEPTION

Year	Source	Type of Wood	Quantum (mt)
1984-5	Open Market	E. hybrid	2,407
1985-6	Open Market	E. hybrid	1,139
1986-7	Open Market	E. hybrid	5,598
		Karuvalem	6,993
1987-8	Open Market	E. hybrid	1,944
		Karuvalem	9,139
1988-9	Open Market	E. hybrid	2,407
		Karuvalem	4,001
1989-90	Open Market	Karuvalem	7,976
		E. hybrid	28,231
1990-1	EH/N LC	E. hybrid	2,238
		Karuvalem	10,012
1991-2	EH/N LC	E. hardwood	15,822
1992-3	NLC	E. hybrid	5,372

Source: TNPL.

Board. TNPL has steadily increased its share of captive generation to total power consumption to 84 per cent. This percentage is much lower than that of other mills.

Effluents: The existing treatment plant is being upgraded to improve the quality of the discharged water. To study the long-term effect of the use of treated effluent for irrigation, TNPL has asked the Tamil Nadu Agriculture University to carry out a study. The TNPL chief said that in the last two months there has been a considerable fall in the discharge of treated effluents. It is now being pumped into some 1,500 acres of arid land brought under sugarcane cultivation under a lift irrigation scheme. 'We have been sanctioned power supply for 14 hours a day for the purpose,' said Mr. S.K. Das (TNPL, M.D.), adding that there would be no discharge of treated effluent into the regular canal, if TNPL was assured of round-the-clock power supply (Krishnan 1994). However, the Tamil Nadu Agricultural University professor said this is dangerous as the soil in the area is clayey (Prof. Obliswamy, personal communication).

Diversification: There is a mention of only two diversification plans of TNPL in its annual reports. The first refers to the active consideration TNPL is giving to the setting up of a few ancillary industries related to the Paper Industry. According to TNPL, it made a proposal of setting up a caustic soda plant in the joint sector in association with a private promoter with the approval of the Government of Tamil Nadu, this being in 1991. Another diversification plan was made in 1992 where the *Annual Report* says that the company is proposing to enter the non-conventional energy sector for generation of electricity through a 20 mw Wind Farm Project and 3 Mini Hydel Schemes (7.5 mw) at an outlay of Rs.75 crore (*AR*, 1991, 1992 and *ET*, 13 April 1992).

During our visit (August 1993), the company executives made no reference to the backward linkages which they had proposed two years ago.

Expansion Programme of TNPL

The company in its annual report states that it has commenced the implementation of the expansion project envisaging doubling of the plant capacity from 90,000 tpa to 1,80,000 tpa at an estimated cost of Rs. 475 crore. The project feasibility report is said to have been approved by the Government of Tamil Nadu. The World Bank which granted liberal assistance for its initial project was considering sanctioning of term loans for the expansion scheme under their Renewable Energy Assistance Programme.

All-India financial institutions like the IDBI, IFCI, ICICI, IRBI, UTI, LIC and GIC had came forward to grant term loans to the company for financing its expansion programme (*AR*, 1992; *FE*, 4 June 1991).

During our field trip, the TNPL executives mentioned that they propose to set up mills in Kashipur (Nainital), Maharashtra and Gujarat for newsprint (personal communication, August 1993).

Summary and Conclusion

Setting up of a paper mill heralds development. However, if it is misused it spells a threat to the ecosystem and to life. Tamil Nadu was keen to have its own paper mills. A forest survey was conducted by the Forest Department, and it found that though the stock of bamboo was much less than expected, it was still adequate to support a PPI unit.

The Tamil Nadu government sought and found its own entrepreneurs whom it strengthened first with financial resources, then with bamboo, later with eucalyptus, and even a legislation which made it possible for Seshasayee Paper Boards to grow its own supply of bagasse through a tripartite agreement with the sugar mills and the farmers.

Tamil Nadu's prejudice against Marwari capital was so intense that it was blind to the danger of opening its doors either to an Italian firm to set up a viscose rayon plant or to the World Bank.

The first mill Seshasayee (SPM), was forward looking. It proposed to set up a bagasse-based newsprint (NP) mill. NP is shunned by the private sector as investments are high and the returns low. SPB approached the state government for help. When it was unable to provide resources SPB turned to the World Bank (WB). WB willingly agreed and stole a march over the parent company and disowned any relationship with it.

NP industry in India is either in the public or in the joint sector. It is in the doldrums, due to the Government of India's short-sighted policy of permitting entry of newsprint from the West into the country without excise, especially when it undercuts its prices to cope with the recession of NP in the west. But TNPL is in no danger. It had the clout to get the Newsprint Act modified. Unlike the other NP units in the country, it is quietly changing its product mix towards production of cultural paper; good cultural paper is the only PPI's commodity in India which is not threatened by the entry of multinational paper giants, provided it is modernized and can match its economy of scale with that of the multinational giants. TNPL with its new expansion is the mill which can, therefore it has an edge over all the other units. Bagasse is a drawback, particularly when the effluents are used to cultivate sugar cane as is done by both SPB and TNPL. Use of effluent spells danger to the ecosystem, and, in the case of Tamil Nadu, even to the social system.

Recommendations

It is imperative that GOI informs TAFCORN, which is struggling to raise *E. tereticornis* to put an immediate stop to it. This information must also reach NABARD. As an alternate in development TAFCORN'S and NABARD's resources should be used to rehabilitate natural forests; and in ecologically appropriate areas raise bamboo for paper and newsprint.

Forest-based viscose rayon production should be banned forthwith. GOI and Government of Tamil Nadu, through proper land legislation should support farmers in growing appropriate crops, which includes pulpable agricultural crops and agave in arid lands. Tamil Nadu is rain-starved and sugar cane is a guzzler of water. Use of effluents is no solution. Its long-term effects on the soil must also be evaluated.

The Ministry of Environment should ban the use of primary-treated effluent for cultivation of cane. It *notes that NEERI* is indulgent towards PPI.

The Supreme Court's ruling (*The Times of India*, 10 April 1996) on closing down all polluting industries, like the tanneries of Tamil Nadu should be made applicable to all industries including powerful TNP and SPB.

NOTE

1. M. Ct. Chidambaram Chettiar, an outstanding banker and industrial entrepreneur who died in a plane crash in 1951, is credited with having started Travancore Rayons sometime after the end of Second World War and before Independence. Perhaps the foremost reason is the remarkable personality of C.P. Ramaswamy Iyer, Dewan of the erstwhile Travancore state, who inspired by the Mysore government, was keen to

set up new industries. The promise of state-aid in getting electricity, cheap land, access to water and forests was vital in supporting the entrepreneurs in their ventures (Baldwin 1959). But the report in the late 1970s gives an impression that rayon in Kerala state was eventually being produced from cotton lintas pulp (Goel undated: 130, in Ahuja).

REFERENCES

Ahuja, S.P. (n.d.), *Paper Industry in India*, the Institute of Market Research, New Delhi.

Andiappan, K. and J. Wilson 1963, 'Bamboos in the Madras State in India', *Indian Forester*, vol. 89, no. 1, p. 259.

Anon. 1985, 'Bagasse-Based Unit of Newsprint Ready', *The Times of India*, 1 July.

——— 1985, '75 per cent Capacity Used by Pugalur Paper Unit', *Economic Times*, 2 August.

——— 1986, 'TNPL Poised for Sustained Growth', *Financial Express*, 23 May.

——— 1991, 'TNPL to Double Output', *Financial Express*, 4 June.

——— 1991, 'SPB Your Spring-Board to Wealth', *Financial Express*, 23 September.

——— 1991, 'SPB Making it Worth the Paper', *The Times of India*, 12 December.

——— 1991, 'TNPL to Double Output', *Financial Express*, 4 June.

——— 1992, 'TNPL to Diversify into Power Sector', *Economic Times*, 13 April.

——— 1993, *Journal of Industry and Commerce*, vol. XXX, May 1993.

Baldwin, G.I. 1959, 'Industrial Growth in South India', *The Free Press*, Glencoe, Illinois.

Baliga, B.S. 1966, *Madras District Gazetteer, Coimbatore*, Government Press, Madras.

Brandis, D. 1883, *Suggestions Regarding Forest Administration in the Madras Presidency*, Government Press, Madras.

ERI Gem 1979, 'Seshasayee Paper', *Financial Express*, 5 February.

FAO 1962, *Economic Availability of Fiberous Raw Material for Pulping in India, Burma and Thailand and the Far East*, vol. II.

——— 1980, *Pulping and Paper Making Properties of Fast Growing Plantation Wood Species*, vols. I, II, Forestry Paper 19, Rome.

Francis, W. 1907, *Madras District Gazetteer, Coimbatore*, Government Press, Madras.

——— 1908, *Madras District Gazetteer, The Nilgiris*, Government Press, Madras.

FRI 1961, *100 Years of Indian Forestry, 1861-1961*, vol. I, Forest Research Institute, Dehradun.

Govenment of India 1908, *Imperial Gazetteer of India Provincial Series—Madras* vol. 2, Calcutta.

Hazare, R.K. 1967, *The Structure of Private Corporate Sector*, Asia Publishing House, Bombay.

Innes, C.A. 1905, *Madras District Gazetteer, Coimbatore*, Government Press, Madras.

Joseph, John J. 1959, *A Report on the Survey of Bamboos Areas in the Madras State,* Forest Department Madras.

Krishnan, G.V. 1994, 'Green Activists Gunning for TNPL', *The Times of India,* 9 April.

Lock, G.W. 1962, *Sisal,* Longmans, London.

Meera, Shenoy 1992, 'SPB—The Best is Yet to Come', *Business India*, 13-26 April.

Mudar, Pathreya 1991, 'Seshasayee Paper Boards, Making it Worth the Paper', *The Times of India*, 12 November.

Pani, Dharani K. 1990, 'TNPL—Trading on the Turnaround', *Business India*, 30 April-13 May.

Raju, Sinha 1991, Seshasayee Fails to Take Advantage of Increasing Demand, *Business & Political Observer*, 21 April.

Rao, Y.A., *The Paper Industry in India—Status and Prospects,* OUP, New Delhi.

Shankarnarayan, K.A. and R.K. Gupta, 1959, 'Vegetation of Coimbatore District', *Indian Forester*, vol. 85, no. 9, p. 538.

Seshasayee Paper Boards, *Annual Reports from 1985-6 to 1992-3*, Erode.

SIDA, 1982, *Social Forestry*.

SPB, 1976, Chairman's speech, *Hindu,* 28 September.

——— 1979, Chairman's speech, *The Times of India*, 25 September.

Stebbings, E.P., 1921, *Forests of India*, vol. I, John Lane and Brodle Head Ltd., London.

——— 1923, *Forests of India*, vol. II, John Lane and Brodle Head Ltd., London.

——— 1926, *Forests of India*, vol. III, John Lane and Brodle Head Ltd., London.

Stock Exchange Official Directory, *Papers on SPB and TNPL*, Bombay.

TAFCORN, *Annual Reports from the years 1970-1973*, Irichy.

TNPL, 1993, *A Brief History of the Industry Research and Development Activities* (a handout), Pungalur.

——— *Annual Reports from 1989-90 to 1992-3*, Pungalur.

———. *A Brief History of the Industry* (a handout), Pungalur.

——— *Research and Development Activities* (a handout), Pungalur.

CHAPTER 9

North-East

Why Hindustan Paper Corporation and How Come Bamboo Forests?

Green bamboo, golden bamboo,
Tall bamboo, giant bamboo,
Stunted bamboo, sprawling bamboo,
Runner bamboo, thorny bamboo,
Bamboo, bamboo everywhere,
It's the enchanted land of bamboo.
Where,
The prowling Tiger fears to tread,
The cunning Jackal doesn't dare to peep.
Where the grizzly Bear having had his fill
Doesn't care.
There are babes in these woods,
It's the enchanted land of bamboo.
You proud senile Lion
Don't heed the foolish Wolf pack
And put your little ones to sleep.
For the law of the jungle is broken,
The hunters are out to kill
And the killers must have their prey
Even if they are only babes.
The hunter will fell the bamboos
To find your babes slumbering.
Where's your pride, old Lion?
Wake up, wake up—
For it's your babes who are in the woods
Oh! Don't put your babes to sleep.

M.S.

SECTION 1

Political Economy of Underdevelopment: Assam and Nagaland

The late emergence of Hindustan Paper Corporation (HPC) and its location in the turbulent north-east raises many questions which need to be answered before one can make a judgement on the apparent malfunctioning of this modern public sector corporation. Only history will tell us why the north-east is underdeveloped and turbulent as well.

Early History

Historical records indicate that since the thirteenth century immigrant population of different races, culture and religion, which included Muslims and Buddhists besides the Hindus had settled in Assam and integrated themselves into that society and even adopted the Ahamiya language. This process continued by and large into the British period. When the British entered the thickly wooded 'Ahom', that is Assam, in 1826 it covered only the Brahmaputra region and a small part of Karbi-Anglong.

Impediments to Development

For sustained economic development a long period of peace and political stability is necessary. But Assam was beset with political turmoil due to frequent civil wars and foreign invasion. The Moamaria rebellion of 1770 and the Burmese invasion of 1817-20 led to total chaos and persecution. Many people perished or fled, and the outcome was a disastrous depopulation. Restoration of the Ahom monarchy in 1833 by the British aggravated conditions of anarchy and misrule. Modern bourgeois development was hampered as the British kept the boundaries of the state in a flux. Right from 1826 to 1873, and again from 1905 to 1921 Assam was tagged on to East Bengal and Bengali was imposed as the official language. This put the local people at a disadvantage, as historically the Bengalis were more advanced in education, trade and business and therefore secured a lion's share of government jobs as 'baboos'.

The Language Question and Underdevelopment

The British had entered the unknown densely wooded region warily, and for administrative convenience tagged on a couple of districts of Bengal to the new province; the most significant of these for our present concern, were Sylhet and Cachar. The British chose also to use the more familiar Bengali language as the official language in Assam. Although Assam had a rich tradition of writing *Buranjis* since 1228, they had no experience of keeping records of land, revenue, etc., which the British administrators needed. Rather than train the Asamiya, the British found it more convenient to fill in the vacancies with 'baboos' from Bengal which had a surplus of educated unemployed.

As Hussain points out, suppressing a native language by a sister language right from 1837-74, created conditions of conflict between the Asamiya and Bengalis, a conflict useful to the rulers to divide the subjugated people. The differences between the two were highlighted deliberately by the provocative language used by the British census officer Mullan in the 1931 census report, when the anti-colonial struggle had reached a high pitch (M. Hussain 1993: 38, 206-8).

The integrative process, whether Hindu or Muslim, of Barak Valley with Assam is on, but attempts to create a divide is part of the political game. The people of Cachar are of Sylhetti stock (Gupta 1984), they are Bengali-speaking and are deservedly proud of their mother tongue. But fear of Bengalis and therefore of the Bengali-speaking people of the Barak valley was politically induced into the Asamiyas of the Brahmaputra region. The Barak valley is now a part of Assam, cut off from the rest of India by Bangladesh. The link with Assam is *a fait accompli* and there is little that the people of the Barak valley can do.

Underdevelopment and Regional Divide

At the same time and for the same purpose, the dubious question of the Hindu-Muslim divide was highlighted by Mullan. One has to go back in time to understand the real situation. As everywhere else, in Assam, too the British monetized the economy and made land a commodity. The British were interested in increasing land revenue. Assam is fertile and had abundance of land but with a low density population. The position of the landed aristocracy worsened as a result of the new revenue system combined with abolition of slavery in 1843. Their estates became unproductive and the Ahom aristocracy was unable to pay even the revenue. The British therefore directly patronized migration of the dispossessed peasantry from East Bengal mainly Mymensingh, Rangpur and Parbani. Sylhet had already been made a part of Assam. The main motive was to generate more land revenue. With the freedom movement picking up, it was politically useful to divide and rule as is evident from the communal colouring given by the Census Commissioner, Mullan. This despite the fact that the peasantry was hardworking, in strong contrast to the locals (Gupta 1984: 116) and integrated itself with the local culture and people and is referred to as Na Asamiya. Mullan's game paid off much later, and is oft quoted by the All Assam Student Union (AASU) and the All-Assam Gana Sangram Parishad (AAGSP). The RSS, and the BIP stoked communal fires while the Marwari business classes and the upper class of Assam provided the resources. This is understandable, as in 1974-5 the CPI(M) and its labour wing was gaining ground in Assam's urban centres and appeared as a threat to the centrist parties (Hussain 1993: 38, 206-8).

Why No Assamese Entrepreneur

The next question is why did Assam not promote a local entrepreneur as Tamil Nadu and Andhra Pradesh had done? Was there no potential business talent? History points to a contrary trend. Nagaland had tried in the late 1960s and failed. The privileged classes of Assam hankered for white-collar jobs. Commercial and industrial pursuits ranked low in their social estimate. The lower classes were influenced by the trend set by the *bhadralok* (the upper class). Meanwhile, the once hardy people of Assam were getting weak, indolent, degenerate, and becoming addicted to opium. Opium cultivation was introduced in Assam in 1770. It was the greatest source of excise for the British. Also, Assam did not throw up any enterprising entrepreneur. Historically viewed it is either the artisans or the traders who turn to factory production of the goods they deal with. But this did not happen in Assam and there are several reasons for this. Assam was more or less a subsistence economy, with fertile land and a predominant primary sector. Agriculture provided the essentials and the household industries provided the domestic needs. Even weaving of 'muga' silk saris was a household activity, and there were no professional castes. There was an acute shortage of coins, and trade was often in the form of barter (Kurshid Alam 1983: 91-2). Such conditions precluded the growth of a powerful trading class.

Immediately after the establishment of rule, there was a sharp decline in trade. The Barapetas and Sualkachis were only petty traders who dealt in articles such as mustard and handicrafts. In the British period they could not withstand competition with the Marwari traders who traded in all commodities and who had over the years in collaboration with the British entrenched themselves at the top of the economy (Alam 1983: 91-103). The British started the tea industry in 1832 and retained its monopoly till the turn of the century. A dozen or two Asamiyas tried to enter the tea industry but remained small-time (Alam). It was the most powerful of the Calcutta-based businessmen, Birla, Dalmia, Bajoria, Bangur, Goenka and later Khaitan who began to buy up tea estates from the departing British planters in the 1950s. Four of them were also well entrenched in PPI, and FAO noted this too. Among the various new locations in India FAO suggested Guwahati in Assam using bamboo and eventually *Pinus khasya* (p. 91) for paper. It also argued: exploitation of tropical forests for pulp alone is not the complete answer, it has to be combined with other forest products, fibre board, saw mills, plywood with high density wood (FAO 1954: 61). FAO's prediction has come true, including plantations of *Pinus khasya*. The PPI giants ignored FAO's suggestion to set up a PPI unit, but gratefully took on the new timber business. Asamiya businessmen hardly exceeded half a dozen and are ranked only as middle bourgeoisie. Apart from this, the middle class is made up of professional and small landlords (Alam) who are in no position to take on the role of entrepreneurs in a heavy investment industry. It was stated earlier that the big Marwari interest is in plywood and saw mills, and the most powerful in the region is no other than Birla.

Theoretically, industry gives a boost to development and the north-east was disgruntled that it was backward. Apart from oil, there was no modern industrial development. Even so, enormous amounts of crude were chosen to be refined in Barauni in Bihar, despite the Indian and Soviet experts' advice. The Noonmati refinery near Guwahati is a mere toy, a sop, to placate the locals (Gupta 1984: 66-7). And modern paper units were expected to be another.

The north-east is not an integrated whole. It is a sensitive frontier area which shares a common border with other countries. The British administrators-cum-anthropologists had pictured the hill people as fossilized segments of humanity. They had failed to recognize several factors such as their growing knowledge of private property rights in stocks of grain, livestock, orchards and paddyfields; the evolution of chiefdoms into statehood; spread of new world crops like chillies, pineapple and wet rice cultivation well before the coming of the British, and the development of a link language, the Nagamese (Guha 1981: 7).

The contradictions within the region sharpened further at the time of partition. The Assamese were in the majority and the power to rule went to them. Like the British, the Assamese were indifferent to the claims of the hill people, to their share in the fruits of independence. The hill movements began, but Nehru initially rejected them. Assam still has acrimonious border disputes with all the states and union territories carved out of it. The bloodiest is with Nagaland where the two often clash near Shibsagar district. Nagaland claims nine reserve forests and some tea gardens measuring 4,975 sq. km now in Assam. Mizoram claims massive chunks of Cachar forests (Gupta 1984: 77). The problems with only these two states are raised as they are relevant from the standpoint of functioning of the three HPC paper units.

But no private capital flowed into Assam to set up a PPI unit.[1] On the contrary, an immense amount of bamboo flowed out to various distant PPI units as and when necessary to fill the gaps during paper famines. Distance was not a deterrent factor. Two units were set up in the deep south as they fulfilled other basic requirements. Birla was already entrenched in the area, his interest was initially timber for tea chests and later veneer which is a qualitatively different proposition from setting up a PPI unit.

The Politics of Nagaland

Nagaland is a small compact state formed in 1962. But the Nagas continue to also live in the border areas of Assam, Manipur and Arunachal Pradesh. In fact their original homeland spills over into the present Burma. Racially they are classed as Indo-Mongoloid.

Not much is known about their history, the anthropological account of administrators fails to note certain basic changes like transformation from chieftainship to state, the emergence of private property, control over method of agricultural production without which their capacity to put up a joint sector heavy investment paper industry is unthinkable. British anthropologists mystified the Nagas as head-hunters and romanticized them as naked and practitioners of dormitory living for the young unmarried. They also failed to tell us that Nagas resisted the British right from the time when their troops entered the Naga area in 1834, till the turn of the century, or how the British favoured the Kukris, the Manipur state and the Hundi princes of the plains against the Nagas. However, it is also true that the British left intact the customs and traditions of the Nagas, particularly of the non-Christian Nagas, as well as the democratic structure of their village life, administration, judicial structures, and village councils (Kuithai and Haksar 1984: 71). This is important in order to understand why the bamboo forests of Nagaland are so full and in such glowing health. The British kept the two social systems, viz., the plains and the hills under different systems of administration and in 1873, passed an Inner Line Regulation to prevent non-Nagas from entering Nagaland and acquiring property (Kuithai and Haksar: 71).

The Naga hills were established as an administrative district in 1881 with Kohima as the centre. In 1888, a new subcentre was opened in Mokokchung among the Aos. The First World War had no impact on the Nagas, but the Second World War proved devastating. A little earlier in 1937, Nagaland districts along with the Lushai hills and the north Cachar hills were decared Excluded Areas within the province of Assam. This meant that no act of the central or provincial legislature could apply to the Naga hills district, unless the Governor of Assam so directed. The Governor was empowered to make regulations for peace and good government.

In February 1946, the Naga National Council (NNC) was formed according to Naga traditions and conventions. NNC's plan was to provide a 'Protected Status' for Nagaland with India as a guardian power for an interim period of ten years. After this, it would decide its political future. In June 1947, a nine-point agreement was signed between the Governor of Assam, Akbar Hydari, and the NNC. The NNC was given recognition as the sole national political authority of the Nagas. In July that year, the Indian Constituent Assembly bypassed the provisions of this agreement. True, M.K. Gandhi intervened and assured the Nagas of their right to self-determination. But the disappointed NNC went ahead and declared complete independence of the Naga people. Tensions mounted and in June 1948, the Governor and the Chief Minister of Assam gave a written assurance that there was never any question of non-implementation of the nine-point agreement. Yet, in July the same year, NNC leaders including Phizo, were imprisoned. C. Rajagopalachari's reassurance to the NNC delegation in July 1949 did not settle the issue. In May 1951, the NNC held a plebiscite and it got the absolute support of the Nagas to remain separate from India, but the Government of India ignored the people's verdict. The Nagas in turn boycotted the first general elections in 1952 and gave a call for a civil disobedience movement. The Government of India reacted negatively, and Nehru's visit to Nagaland in March 1953 was a total failure.

For whatever reasons, perhaps as a retribution, the Inner Line Regulation (1873) was annulled in 1953 itself and businessmen (by and large, Calcutta-based Marwari veneer and timber manufacturers) entered, mainly to get hold of its timber. As a result the face of the Nagaland forests has been completely transformed. Battalions of armed forces also entered and military control tightened. The year 1954 witnessed the Chingemi massacre, and in 1955 the Assam Disturbed Areas Act was passed. The missionaries who had worked with the Nagas, and made them literate were forced to leave after 120 years. It was in the Ao country that the British missionaries had set up their headquarters. In 1956, army units occupied Kohima and other towns, and NNC in March that year inaugurated the Federal Government of Nagaland, raised a standing army and National Home Guards.

Recognizing the determination and solidarity of the Nagas behind its federal government, an I.B. officer, Dutt, in 1957, granted autonomous district status to Nagaland and Tuensang Areas under a commissioner. But he also formed the National People's Convention of handpicked Nagas. In 1958 Nagaland came under army rule, army powers always exceeding the constitutional provision and alienating the people even more (Luilkui and Hanson 1984). The NPC held its second convention in Ungama in 1958 in Mokokchung district and the third in 1959 in Mokokchung in Ao area (Elvin 1962: 64-9). Finally, in 1962, the state of Nagaland was formed and placed under the Ministry of External Affairs. State repression continued. Soon after the third convention, an NPC delegation met Prime Minister Nehru in July 1960. In Parliament, Nehru promptly announced his decision to set up the Nagaland State, which was inaugurated on 13 December 1963 (Das 1982: 274).

It is interesting to note that the Nagaland Paper Mill was conceived, and almost inevitably, was situated in Tuli in Mokokchung district, an Ao area. The paper mill is said to be the brainchild of Shashimaran Aier, a local leader. The Government of India's decision to locate it in Tuli rather than in Naginmora (a better location), seems to have been to win the loyalty of the Aos; also it cannot be a coincidence that the chief executives of the mill till very recently were Aos, and incidentally so too was the Principal Chief Conservator of Forests of Nagaland. This brief history of the area is necessary to understand the history of its forests.

SECTION 2

Natural History of Forests of North-East

The most striking feature of the forests of the north-east are bamboos—lush, full and strikingly beautiful. It could not have always been a bamboo forest, the relief and climatic conditions do not warrant this. Geographical conditions point to a much more evolved and varied complex forest ecosystem; whereas from the evolutionary point of view, bamboos that are found to dominate the scenario are on a simpler plane, belonging to the natural order of *Graminea.*

The grasses have evolved into numerous tall woody species, primarily in response to human interference, but the form that they have taken is the result of the natural geographical conditions. It is therefore necessary to look at both ecosystems and human intervention to appreciate the natural history of forests in the north-east. Our observations are largely confirmed to the region around and along the route to Guwahati, Nowgong, parts of Mokokchung district of Nagaland and from there to Cachar, Karimganj, and back to Guwahati, via the Khasi hills.

In this section it is necessary to first present the human interference in these forest areas which changed the course of the natural history of the forest as recorded at the turn of the century.

Assam

The British took control of Assam in 1826. A British civilian discovered a tea bush in 1832, an exciting discovery, as tea was a precious exotic commodity in Europe. Within two years its plantation activity began. The best of the forests were cleared for tea, and forest reserves were pushed into the interior regions. In upper Assam timber exploitation came up only with tea cultivation to make tea chests (Stracey 1954: 759).

In 1852 a system of lease was introduced in Goalpara and Kamrup districts for the extraction of forest produce. There was no serious forest conservation till 1862, till Brandis, the first Inspector General of Forests, visited Assam. Between 1861 and 1878, 671 sq. miles were marked off for reserve forests. This was raised to 9,539 sq. miles in 1884. Forest not declared as reserved came under the Revenue Department. The Forest Department of Assam came into being only in 1874-5. Meanwhile, from 1864 onwards, Dr. Anderson of the Botanical Gardens of Calcutta was considered the chief conservator of forests. During his stewardship the first teak plantations were tried out in Lakhimpur district in 1869 and in Kamrup in March 1872 (Stracey 1956: 619). Even at this time it was still considered, more important to extract forests rather than generate them. In Assam, standing trees in the forest were sold to contractors who would then take them down by animals or float them down the rivers. Sale of timber was profitable to the British. This started first in the Surma or Barak valley and spread into the Brahmaputra valley again only with the advent of the tea industry (Stracey). The demand for tea chests led to the springing up of a number of saw mills in 1884.

In 1894, restrictions were brought down on *jhum* cultivation especially in timber catchment areas. In 1886 the *taungya* system, a combination of agriculture and tree plantation was introduced to generate plantations, as timber trees which were used for tea chests were fast vanishing.

In 1880, collecting wild rubber became a feature of the Forest Department (FD) activity. But the stock began declining and in 1893, 2,472 acres of rubber saplings were planted. Plantation activity was stopped before the end of the decade by the orders of the Chief Commissioner who wanted efforts to be concentrated on improving the natural forests; it was revived only in 1921. Earlier, in 1911-12, Pearson, the forest economist toured the area to enquire into the state of the tea-box industry, while Raitt investigated the suitability of bamboo for paper in 1912. A silviculturist was appointed in 1927 and the post was made permanent only in 1946 (Stracey: 620-1). All this no doubt indicated an exhaustion of timber trees and a concern to save the forests.

Earlier, with the advent of railways in the state, much of *sal* and other local species had been cut. Simple working plan surveys commenced in 1889 and ceased in 1895. Only *sal* was considered valuable (Stracey: 621). In the second decade of this century, tea boxes began to be made by two British-owned mills which were opened in 1921 and 1924. A forest utilization officer was appointed in 1935

(Stracey: 620) to cope with the emerging demand. The forest department was busy during the Second World War. Four American land lease saw mills were set up in various parts of the state for rapid production of sawn timber for the Second World War. Rs. 4.3 crore of timber, Rs.13.30 crore of bamboos and thatch grass were supplied to the army between 1942 and 1945. Due to this there was a diminition of first class timber; introduction of secondary species was the result of the war (Stracey).

Plunder did not cease, Stracey reports that there were still areas of idle forests particularly in the central zone (pp. 621-2). This perhaps is why others entered. Earlier we had mentioned that powerful Calcutta based Marwaris had bought over tea estates from English planters who chose to go back to their home after the transfer of power. The planters needed thin ply chests to pack tea. As in Kerala (see Ch. 7), so here too, from tea chest to plywood and veneer was only a profitable step forward for the Marwari businessman. And in this it was guided by FAO (1954: 61) with the blessings of GOI which was possibly unaware of the ecological damage. By 1956, there were as many as 20 veneer mills in Assam which used up as much as 16,000 tpa of wood. In its pre-investment survey, FAO (1970), also suggested conversion from axe to saw logging and improving road and river transport. This demand saw many of the tracts in the interior being opened out for exploitation. There was a boom in timber trade and large timber trees were being wiped out, transforming the north-east forests and converting many parts of it predominantly to bamboo. Both the *Imperial* (1908) and the *District Gazetteers* (1905, 1911) mention bamboo as one of the prevalent species in the mixed forests, whatever the variety, without at any point giving an impression that there were pure bamboo brakes anywhere in the north-east at the turn of the century. The old Gazetteers which describe as well as chronicle events, are useful references to assess the quality of the change.

Physiography

Relief, drainage, rainfall, temperature, soil and its mineral content determine the forest type of a region.

Relief: The river Brahmaputra has carved out its own valley between the two parallel ranges, the Assam Himalayas (north) and the central Assam range or the Shillong plateau. The river runs 720 km crossing every district of Assam.

To the south of the Shillong plateau there is a smaller river valley—the Barak valley. This valley is enclosed by ranges on all three sides. Cachar is the only district in this valley.

The northern hills and the Assam Himalayas had formed a part of the sea of Tethys which separated it from the Gondwana land of the south. During the tertiary age, owing to the crustal movement of the earth, the seabed rose and was folded into a series of mountain chains. The outcrop of the red earth deposited at the foot of the hills may be sediments from this seabed. The Shillong plateau is considered to be a continuation of Gondwana land. The Brahmaputra and Barak valleys are alluvial in origin.

Based on its physical features Assam may be divided into:

(1) Assam Himalayas,
(2) Brahmaputra valley,
(3) Shillong plateau,
(4) Barak valley, and
(5) South-eastern Hilly Region.

Major hill ranges of the state are a direct consequence of the geological features. Geology can influence the vegetation which may change due to a variety of pressures including biotic pressures.

(1) The Eastern Himalayas rise rapidly from the plains unlike the Western Himalayas. These mighty mountains have grassy expanses intersected by rivers.
(2) The Brahmaputra valley can be divided into the lower valley, where the plains on either side of the river are broken by hills, and the upper valley where plains lie uninterrupted from the Himalayas to the Naga hills, i.e. from the north-west to the south-east. The plains have been formed by the alluvium carried down by the Brahmaputra. Originally the area must have had luxuriant vegetation but all of it was covered by mud and in course of time emerged as the coal and oil reserves of the upper valley.
(3) The Shillong plateau is really a range of mountains which acts as the water-shed between the Brahmaputra and Barak valleys. In the west it consists of some broken hills, in the east it forms a part of the north Cachar hills and then joins the Naga hills. The plateau consists of grass downs and rivulets. Enormous rainfall has hollowed the river valleys into ravines with luxuriant vegetation. The southern face of the plateau facing the Barak valley is abrupt while the northern edge slopes gently to the Brahmaputra valley.

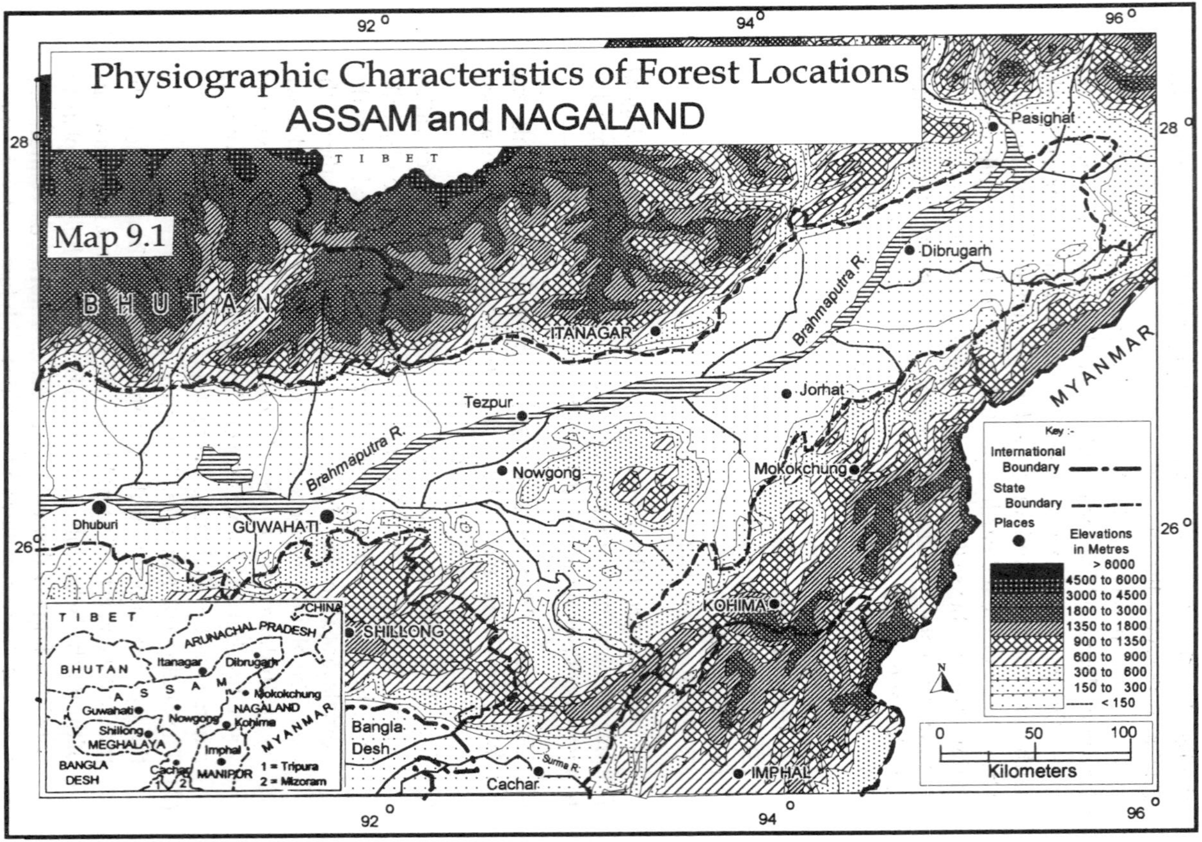
Physiographic Characteristics of Forest Locations
ASSAM and NAGALAND
Map 9.1
92°
94°
96°
28°
26°
TIBET
BHUTAN
ITANAGAR
Pasighat
Dibrugarh
Brahmaputra R.
Jorhat
Tezpur
Nowgong
Mokokchung
Dhuburi
GUWAHATI
SHILLONG
KOHIMA
IMPHAL
Bangla Desh
Cachar
Surma R.
MYANMAR
CHINA
ARUNACHAL PRADESH
Itanagar
ASSAM
Guwahati
Shillong
MEGHALAYA
NAGALAND
Kohima
Imphal
MANIPUR
BANGLA DESH
1 = Tripura
2 = Mizoram
Key :-
International Boundary
State Boundary
Places
Elevations in Metres
> 6000
4500 to 6000
3000 to 4500
1800 to 3000
1350 to 1800
900 to 1350
600 to 900
300 to 600
150 to 300
< 150
N
0
50
100
Kilometers

(4) The Barak valley consists of swampy flats with many hills and ridges. The Garo hills mark the northern limit and the south-eastern boundary consists of long spurs of highland projecting from Tripura and the Mizo hills. The valley constitutes a compact group of hills covered by jungles.

(5) This belt includes the Naga, Mizo and Manipur hills. Due to a number of geological strikes and faults, the mountain-building movement of this region continues. The ranges run in a north-south direction and are separated by narrow valleys covered with green vegetation.

Drainage: The major rivers are the Brahmaputra and the Barak or Surma. The Brahmaputra originates in Tibet and flows eastwards as the Tsangpo, cutting through the Himalayas and then turning south-east after which it takes on the name Dihang and is met by another stream, Dibong, from the north. Finally, a third stream, Lohit from the east, joins the two at Brahmakunda and the river then flows as the Brahmaputra. After crossing the end of the Garo hills it turns south and meets the Ganges at Goalundo. The Brahmaputra travels through Assam for 720 km. The river has a large number of tributaries.

The Surma is an important river on the south of the Shillong plateau. It rises on the Barail range, in Manipur, flows through Cachar and then empties itself into the Brahmaputra. The length of the river is 901 km from source to mouth and it receives many tributaries.

Rainfall: Assam receives very high rainfall which is concentrated in the monsoon months from June to September. There are occasional thunder showers in the pre-monsoon period. It is a typically tropical moist climate. A few showers are received in late October and early November from the retreating monsoons.

Forests Types—Assam (as classified by Seth & Champion)

(1) Tropical Evergreen
(2) *Sal* Forests
(3) Mixed Deciduous Forests
(4) Riverine Forests
(5) Pine Forests
(6) Savannah Forests.

(1) The Tropical Evergreen species of *Messua ferrea, Shorea assamica, Dipterocarpus macrocarpus* are the main types. In the plains of Upper

Assam there are large patches of evergreen forests between the tea and paddy cultivations. In Cachar this type of forest is confined to the north and the east of the lower slopes.

(2) The home of *Shorea robusta*, *sal* forests lie in the districts of Kamrup, Goalpara and Garo hills. Darrang, Nowgong and Khasi-Jaintia hills also have small areas of *sal*.

(3) Mixed Deciduous—This type is dominant in lower Assam along the drier slopes and ridges of hills.

(4) Riverine Forests—The dominant species, *Acacia catechu and Dalbergia sissoo*, are along the main river banks, often pure and sometimes mixed. Another type of riverine deciduous forest occurs in areas that have flowing streams even in summer. The dominant species are *Bombao malabaricum*, *Albizzia procera* and *Anthocephalus cadamba*.

(5) Pure forests are confined to Shillong and Khasi-Jaintia hills. Today, Caribbean pine plantations are very common.

(6) Savannah occurs in well-drained highlands. Lowland savannahs are located on the banks of rivers and streams.

Anthropogenic interference has brought about certain changes. S. Rajkhowa (1961: 520-32), has some difficulties with this classification. Our concern is with the prevalence, in fact, dominance of bamboo as the major species today. Assam is also rich in bamboo which is found mainly in the first four areas.

The most characteristic feature of secondary growth in both evergreen and semi-evergreen forest areas is bamboo, either as an understorey or as mainly bamboo with a few scattered timber trees. The characteristic species of bamboo found on loamy soil are *Pseudostachyum polymorpha* and *Dendrocalamus hamiltoni* and in the laterite soils, *Melaconna bambusoid. D. hamiltoni* occur largely in low hills of upper Assam, and *Melacoids* in Cachar and the Garo hills. In the drier regions of north Cachar district *Bambusa tulda* is the dominant species (Rajkhowa 1961). Today *B. tulda* grows well as homestead bamboo in the Nowgong district. The bamboo forests tend to become more and more pure as the intensity of shifting cultivation increases, and especially in lateritic soils as in Badshahi *tilla* reserves of the Cachar district. Plantations of timber trees are difficult to maintain as they are soon swamped by bamboo. Where bamboo is not prolific, *Duabanga sonneratiodes* as a pure form takes over on hill sides and *Tetrameles nudiflora* near the streams (Rajkhowa: 532).

Status of Forests in Assam

Although the Nowgong district has the least amount of rainfall, it is well-watered by streams and rivers. This region is semi-deciduous, and deciduous forest area, interspersed with grasslands extending from the Lumding valley, the Mikir hills, the north Cachar hills and the south-west plains of the Nowgong district.

The Dhansiri valley is an evergreen forest area. In Kamrup, Goalpara and the Garo hills the vegetation is mostly of the deciduous type. To the east the Sibsagar and Lakhimpur districts are again evergreen vegetation areas.

Early accounts of the forests of the north-eastern region are recorded by the *District Gazetteers* (1911, 1914). According to Dr. Hooker, Khasi flora is the richest in India.

As early as 1906 areas in the Khasi and Jaintia (K & J) hills were already devoid of forest cover. It was then attributed to the system of *jhumming*. The only forests were the sacred groves limited to the higher reaches of the mountains. The southern half of the K & J hills are bare of trees and covered with grass. Open forests of the Khasi pine that were seen mixed with oak have practically vanished. The northern half of the district used to contain large stretches of evergreen forests with *Mesua ferrea*, *Schima wallichi*, *Vatica lanceofolia*, etc. The *sal* forests are found as far east as the Shillong-Guwahati road. The outer ranges of hills on the north are covered with bamboo and other deciduous trees.

It is said that the Naga and Manipur hills were covered with evergreen forests though from most of the hill sides 'the trees have vanished before the cultivator. The tree species found here were of *Mesua ferrea*, *Lagerstraomia species*, *Michelia*, *Artocarpus*, *Cedrela toona*, *Dipterocarpus species*, etc.

The districts of Cachar and Kamrup had good timber-yielding forests which were marked as reserves. As late as 1906, B.C. Allen reports that to the south much of the country was covered with primeval forests. Further north, higher lands were covered and planted with tea. The hills of Cachar were covered with mixed deciduous and evergreen forests. Where the forests have been *jhummed*, trees have given place to bamboo. Low swampy valleys are covered with cane and reed. The Cachar hills had a lot of good timber trees and permits were issued to traders who wished to extract the timber. The hills of north Cachar rise as a rule in steep slopes covered with forests and bamboo jungles (Allen 1906: 6).

Physiography of Nagaland and its Vegetation

The state of Nagaland is divided into three districts—Kohima, Mokokchung and Tuensang, with Manipur to the south, Burma to the east and the Assam plains all along the western foothills. The state is entirely covered with hills that break into a chaotic wilderness of spurs and ridges, or sometimes, as around Kohima, descend with gentler slopes (Elvin, 1961).

Most villages are located at 3,000-4,000 ft elevations. The foothills are only thinly populated.

Rainfall varies from 70 to 100 inch a year. The major rivers are the Dzulu, Dhansiri and Diphu.

The vegetation is classified as: (a) Sub -Tropical Wet Hill Vegetation: Found at altitudes between 1,000 to 4,000 ft characterized by species of chestnut, *Michelia*, *Gmelina arborea*, *Albizzia*, etc. Pine trees associated with *Rhododendron* and oak are found in the Chakesang region. (b) Wet Temperate Forests: Found at altitudes above 5,000 ft they are characterized by *Betula*, *Rhododendron*, *Magnolia*, *Runus*, *Juglans*, etc.

The vegetation on the lower hills closer to the roads is subject to *jhum* and terracing, and has been stripped of its old vegetation. The *jhum* lands generally get converted into grasslands and the soil gets replenished when bamboo takes over (Ramakrishna).

Bamboo Forests of Nagaland

It seems from the early accounts of the flora that bamboo was found growing mainly among the deciduous trees. While the colonial foresters blame the *jhumming* practice as being responsible for the rapid growth of bamboo, it could also be the large-scale extraction of the deciduous species (good timber species) which opened the canopy and created space for bamboo to flourish. Bamboo is now dominant in these hills. The old accounts do mention bamboo and the numerous species but not as being dominant. The north-eastern region today stands as a bamboo forest area with other forests like evergreen type, deciduous type, *sal* type, etc., to be found only in vestigial patches. Our observations conform to this.

In Nagaland the people reported how once this very same bamboo used to grow only along the river banks, but now has started conquering the hill slopes.

S.N. Rai (1989) makes an interesting observation that in the north-eastern region when the tropical rain forests of the lower elevations get cleared, bamboos take over as colonizer leading to bamboo brakes. He goes on to say that there was heavy exploitation of species like *Terminalia myriocarpa* and *Dipterocarpus macrocarpus*, the former for teachests and the latter for the plywood industry. Even the accounts of the early Gazetteers (Refer to Section on 'Past Status of the Forests') highlight the fact that the very first intruders into the natural forests were the makers of tea-chests. The practice of *jhumming* was labelled by the British and by Indian foresters, except the discerning ones, as the prime cause for the opening of the forests to plant colonizers. This is merely a half-truth. In Assam the *jhummers* had access only to forests on the lower elevations and to the moist deciduous patches. Besides, *jhumming* was an ancient practice which had co-existed with the timber forests. This study has discussed earlier how large clearance of forests for timber or tea began only in the nineteenth century and was initiated by the British in Assam.

Granted, the causes for the abundance of bamboos may be many. Bamboo is an aggressive grabber of space. It was observed in Karimganj district (the catchment area of HPM, Cachar Paper Mills) that vast hill slopes were taken over by *Melaconna bambusoides.* The bamboo has obviously crept in from the banks of streams. This happened when the slopes were cleared of the tree cover, opening up the area for the bamboo. In the midst of the stands of bamboo, here and there tall trees which had probably escaped the axe stood as testimony of the past vegetation.

The question of *jhumming* in Nagaland is somewhat different. It is a hilly terrain which offers little scope for agriculture other than *jhumming* or the more advanced terrace cultivation. But experts have indicated two important constraints which limit the space for *jhumming*, implying that it is not destructive. A family can clear just about 2 ha of forests. The first difficulty is that a family should clear enough space for subsistence; the girth of the trees and the density of the growth should allow a family not only to cut the trees but also to chop and spread them evenly for burning just before the monsoon. Second, there should be an even layer of soil (Pingle et al. 1984: 145). Nagas did practise *jhumming* and in some areas continue to do so. Since 1983, timber was also being taken away in large quantity by Calcutta-based Marwari business houses. Dense bamboo growth on the hill slopes is evidently a combination of the two factors. Bamboo once established, its smothering habit precludes establishment of seedlings of most other trees, and so helps to form self-perpetuating communities (Witmore 1975: 232). P.S. Ramakrishna points out that bamboo plays an important role in nutrient conservation at 100 m high sites from the tenth to the thirtieth year of fallow re-growth. After destruction, nutrient recycling of the slash-and-burn begins, the system soon recovers through a quick succession, and the elements highly susceptible to leaching and run-off are conserved in the living biomass of the bamboo.

Details of the raw material requirement and supply to the three paper mills is dealt with in the section on raw materials. On the raw material issue the three paper units in this area are at present quite comfortably placed. Will they always be so? They can avail the natural bamboo from their allotted catchment areas as well as from homestead bamboo. The local people have a long tradition of growing bamboo on their lands. Essentially *Bambusa tulda* and sometimes *nutans* are the homestead species while *Dendrocalamus hamiltoni* and *Meleconna bambusoides* are the dominant natural species, the former being common in the Brahmaputra valley and in Nagaland and the latter being common to the Barak valley area.

Management of Bamboo

The management of bamboo resources of the region has to be looked at from the ecological, local and the PPI points of view. Table 9.1 gives the quantum supplied to the mills by Government of Assam from eight

TABLE 9.1: RAW MATERIAL SUPPLIED BY THE FOREST DEPARTMENT TO HPC UNITS IN ASSAM

Year	Division	Allotted quantity (AD mt)	Paper mill
1991-2	Kamrup East Division	3,500	HPC, Nagaon Paper Mills, Jagi Road
1991-2	Nagaon Division	3,500	-do-
1991-2	Nagaon South Division	1,000	-do-
1991-2	Dibrugarh	2,500	-do-
1991-2	Digboi	7,000	-do-
1991-2	Doomdooma	2,500	-do-
		20,000	
1991-2	Silchar Division	35,000	HPC, Cachar Paper Mills, Panchgram
1991-2	Karimganj Division	35,000	-do-

Source: Forest Department Guwahati, Assam.

divisions including Karimganj and Table 9.2 forest areas reserved for supply of bamboo.

The Table 9.3 gives the supply of raw materials to Cachar Mills from various sources.

TABLE 9.2: FOREST AREAS RESERVED FOR HPC UNITS IN ASSAM

Name of the district	Assigned forest area
Nowgong	All reserved forests and unclassed state forests in Nowgong district.
Karbi-Anglong	All forests including the reserved forest under the control of the Council.
Cachar	All reserved forests and unclassed state forests in Cachar district.
Kamrup	All reserved forests and unclassed state forests excluding those already assigned to Ashok Paper Mills.
Dibrugarh	All reserved forests and unclassed state forests for extraction of bamboo only and excluding those required by the Governor to meet local needs.

Source: *Agreement Copy*, 20 July 1977.

TABLE 9.3: QUANTUM OF BAMBOO AND HARDWOOD SUPPLIED SINCE INCEPTION TO CACHAR PAPER MILLS, 1984-5 TO 1992-3

(A) Forest Dept. (N.C. Hills and Barak Valley)

Year	Bamboo (ADmt)	Wood (ADmt)
1984-5	5,309	—
1985-6	41,438	—
1987-8	71,256	—
1988-9	62,551	—
1989-90	53,270	—
1990-1	85,748	2,198
1991-2	81,465	1,183
1992-3	78,174	2,291
1993-4	63,145	2,098

(B) Quantum of Raw Materials Secured from Lease Areas

Lease Areas	Quantum of raw material secured (ADmt)
Barrak Valley	70,000
N.C. Hills	1,00,000

Source: Cachar Paper Mills.

Extraction and Felling Practices, Allotment of Working Circles

The Forest Department and North Cachar Hills District Council are approached every year, for allotment of areas and quantity to be harvested, in connection with the working of bamboo. On receipt of the allotment orders, felling and extraction operations are taken up by HPC through its agents. In order to keep a proper check on the operational activities in totality, the job is supervised by trained forest staff of two HPC mills. Each person is assigned to look after one coupe area. The bamboos are manually felled and transported by truks. Operational activities continue from November to May the next year (HPC, Cachar: Note on raw materials, no. 1993).

The spread of various local species of bamboo as secondary growth is a historical fact. But as an ecologically sound species it needs to be tended and harvested with care.

Industry's 'Management' of Bamboo—Our Observations

Setting up of PPI units in the north-east is a result of a breakdown in the industrial development of India and a sorry makeshift arrangement to produce cultural paper for the poorer sections, and for the government's use. As everywhere else, the paper units in Assam too are allotted lease areas from which they are permitted to draw their raw material according to the lease agreement and like everywhere else they are expected to follow the silvicultural practice of felling. In Assam they are not permitted to engage their own labour to harvest the raw material, but are compelled by HPC to invite tenders from local contractors to provide the specified amount of bamboo from the allotted coupes. Incidentally, the mills have a surplus labour force in other departments but due to bureaucratic controls this force cannot be shifted to the task of felling and collecting the raw material. The contractor who submits the lowest tender is picked. He in turn perhaps allots it to his untrained labour which works fast, indeed making almost a clean sweep of the area.

The catchment area lies all around of the Nowgong mills. The nearest coupe that was worked the previous year was about 20 km away, and the predominant bamboo species here is what the team knows as the tall and stately *Dendrocalamus hamiltoni*. J.S. Gamble (1978:

84) describes it as bamboo with large culms 4 to 7 inches in diameter, 40 to 60 and even up to 80 ft high. Our observation in a one-year old worked area was very different. There were only occasional clumps of *D. hamiltoni*, short (never more than 10-15 ft) and almost sprawling with stems curved downwards. This amazing metamorphosis, we were given to understand, is a reaction to anthropogenic interference and is confirmed by the *D. hamiltoni* clumps found only along the roadside even in Nagaland. This species in Nagaland had flowered gregariously, turned brown and was still standing tall, sometimes slumped and broken in a sea of green *B. tulda*. The local people told us that *D. hamiltoni*, though tall, clean and erect at the base, tends to get its stems at the top tangled, and therefore it is difficult to follow the silvicultural prescriptions laid down in the forest manuals. A chance meeting with a forest contractor of HPC during our visit, gives some clue to where the problem lies. Silviculture was an unfamiliar word, its significance being totally unknown to him. In all earnestness he claimed that he was only expected to deliver a specified quantum of bamboo. The forestry sections of both the mills were unhappy at the arrangement of calling for tenders, but were helpless against the rules set by the corporation. The general manager of raw materials of Cachar Mills, a trained forester, had refused to take delivery of the sub-standard forest raw material brought by the contractor (personal communication, November 1993). He has no control over the bamboo cutters under the present system. *Melaconna bambusoides* as single-stemmed is very easy to cope with. In the observational visit to its catchment area in Karimganj, we saw that it was an area rich in multiplicity of flora but with bamboo as its major species, in perfect health. The supply and types of bamboo for Cachar Paper Mills are represented in Figure 9.1 in air-dry metric tonnes (ADmt).

Local Management of Bamboo

Even if *D. hamiltoni* in the Nowgong catchment area threatens to metamorphose and even die out, the area is still wonderfully rich in homestead bamboo. We were shown areas which belong to the tribals as well as to a private household. The latter had a modest home made of bamboo. The small compound was on the state highway and separated from it only by a fairly deep natural drainage channel about 10 ft wide. It was connected to the main road by a bamboo bridge. The compound had only three well-maintained clumps, two of

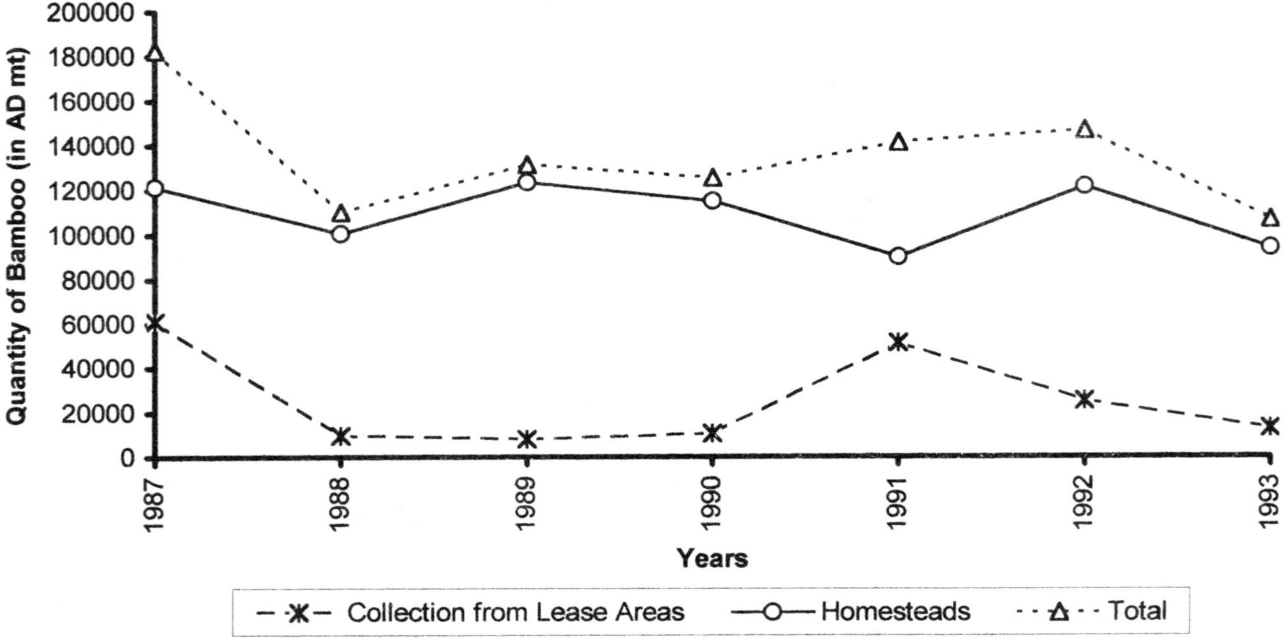

Source: Cachar Paper Mills, November 1993

Figure 9.1: A Comparative Picture of State and Homestead Bamboo used by Cachar Paper Mills

Bambusa tulda and one *Bambusa nutans*. They are hardly distinguishable from each other except for the distance between the nodes/septum. Both the species were tall, erect and uncluttered. The owners had taken care to extract only mature bamboo in the centre of the clump. A large number of mature culms on the periphery were left uncut, which they said would serve as support to young culms from falling over and which also help to keep the rhizome alive and spreading outwards. These families made their livelihood selling bamboo to the local people and to the mills as well.

In Nagaland the people raise bamboo, generally *B. tulda* and some *D. hamiltoni*. These are a breathtaking sight on the steep hill slopes. The Nagas put bamboo to multiple uses. They eat the tender shoots, cook their delicacies in a hollow bamboo stem, build their homes and bridges with bamboo and use the surplus, particularly dead and over-matured bamboo as fuelwood. They sell bamboo to numerous paper mills that ask for it. Yet their bamboo appears to be in perfect health and splendour. In other words, they know how to tend their bamboo. The Forest Department of Nagaland has no right to any forests—forests are owned by the people.

So right within their own north-eastern region is the answer to the sound management of the bamboo resources. The well-tended bamboo 'plantations' of the villagers could be the answer to the industries' demand.

Why Plantations?

In a region so rich in bamboo it is difficult to understand why the industry is demanding land to raise controversial exotics. True, eucalyptus was given as an option to many units in the rest of the country but the status of those plantations even on the Western Ghats is in no way commendable, since none of them is producing anything close to their estimated yields. Belatedly, the Forest Research Institute, Dehradun, has expressed doubts about eucalyptus (Senior Director, personal communication). The raw materials need not be uniform for the whole country, more so, because India offers diverse types of soils, rainfall, topography, etc. The species to be cultivated must be specific to the region and so must also be the management of that material.

Just to emphasize this point we state the example of the north-east. There is an excellent system of management of the bamboo practised by the local people. The industry meets its demand from local varieties as detailed in Tables 9.4 and 9.5. Nevertheless HPC wants land to grow exotics. The Nowgong mill has started its experiment growing *D. strictus* on the advice of its consultant, a retired IFS officer. This is a species which once grew abundantly in the dry deciduous forests in other parts of India and was taken on by paper mills, and therefore believed to be the best pulpable species. But this is not true, says McLure (1966: 64). Its fibre length is shorter than other bamboo species. The fact remains that it does not thrive in moist areas and any attempt to grow *D. strictus* in Assam will definitely not meet the same success as if it were to be raised on its native soil. An attempt to introduce it in evergreen Simlipad forest in Orissa had failed (Sarangi PCCF [retd.], personal communication). *B. nutan* was brought in from Assam and it established itself well in Orissa. The question here is why should every paper mill think in terms of its own plantations? Why not just make use of the efficient system available locally? Maybe a system of acquiring bamboo from the local farmer could be developed through

TABLE 9.4: OTHER SOURCES OF BAMBOO FOR CACHAR MILLS (DIVISION-WISE)

Forest division	Growing stock (ADmt)	Annual availability (ADmt)	Total forest area (ha)	Bamboo area (ha)
1. Kamrup East Divn.	1,66,660	25,000	82,900	42,196
2. Nagaon Divn.	86,650	13,000	1,58,811	20,646
3. Digboi Divn.	1,33,300	20,000		
4. Doomdooma Divn.	33,330		45,283	—
5. Karbi-Anglong East & West	14,20,000		10,33,281	6,43,965

Source: Cachar Paper Mills, November 1993.

TABLE 9.5: CLASSIFICATION OF BAMBOO USED BY CACHAR MILLS

Percentage	Local	Botanical
30	1. Kako	*Dendrocalamus hamiltoni*
30	2. Jati	*Bamboosa tulda*
15	3. Hill Jati	*Oxitenenthera parviflora*
25	4. Bhaluka	*Bamboosa balcooa*
	5. Muli	*Melaconna bambusoides*

Source: Cachar Paper Mills, November 1993.

which both parties would stand to gain. This brings home the fact that the local people, tribal or otherwise, know how to tend their own forests and their local species. It also brings home the fact or the fallacy of introducing exotics on a major scale which does not match the site.

Role of Forest Departments

The major threat to the forests of Assam is from the plywood industry. This is probably true of Nagaland too, after 1953 when the Inner Line Regulation of 1873 was annulled, allowing non-Nagas to enter. The Forest Department of Assam finds itself helpless against the timber and veneer mafia that this industry has encouraged. The mafia operates from outside the state and has the local youth on its side. These unemployed youngsters mobilize local labour to fell and move the wood illegally (PCCF, Assam, private communications). The wealthy who control the timber trade, have links with politicians, a problem that the Forest Department is unable to cope with. Yet another threat to the forests are 'militants' who are rehabilitated by the government. We were informed of this by a retired forester who was upset that in a bid to win them over, the government woos them with promises of land and housing. So forest lands are gifted away to these 'reformed' militants.

In Assam it seems there have always been dense forests which were taken for granted; they are still there, but their quality has changed from a variety of mixed to predominantly bamboo; they were there and they will remain there. Illegal felling needs to be stopped; legal felling was started as late as in 1986. According to a Forest Department communication from the Secretary of Forests to the CCF, the Assam government decided to ban all felling in natural areas. The Forest Department of Assam also mobilized a special squad to police the forests. But how far these orders can be actually implemented, given both the anarchic circumstances in the state, and the Forest Department being limited because of inadequate staff to police the vast expanse of forests, added to which a large number of disgruntled people who are angry with the injustice meted out to them by the state, is the question. Within the state itself there are two districts, viz., Karbi-Anglong and north Cachar hills which are under the administration of the Hill District Council. These areas have prime forest land and the management of these rests with individuals and not with the Forest Department.

In the state of Nagaland there is a Forest Department but this department cannot do much since almost 90 per cent of the land (natural forest areas included) belongs to the people. There are no government lands in Nagaland yet most of the old timber trees have vanished. Manufacturers of veneers and the saw mill owners across the border get hold of them. The most powerful ones are Birla, Goenka, Khaitan, Bangur, Bijoria and Dalmia, all Calcutta-based Marwaris, five of them also being owners of paper mills. As in Assam, so in Nagaland too, it is the unemployed Naga youth which is enticed into illegal felling activity (senior I.F.S. officer from Nagaland cadre now in FRI—Dehradun, personal communication). What is visible is only dense growth of tall *Bambusa tulda* and some *D. hamiltoni* and occasional patches of wild banana. People have tended their bamboo forest with care and they will continue doing so, since it is the only source of sustenance for them.

The Forest Department of Nagaland is trying to promote forestry by supplying seedlings of broad-leaf species to the people which they can plant on their lands after *jhumming*. The people already have well-tended bamboo plantations and also timber species which they sell to outsiders. Of course much of the large trees in the natural forests have been axed. The forests have become a valuable source of money for the people and there is much demand. Any government ban on felling will not apply to the private landholdings. People need to be educated on the need for conservation but alternatives have to be provided to ease the pressure on the natural forests. Supply of seedling of timber species to be raised must be promoted but adequate research must go into the choice of species and their economic and ecological viability. Handing over forests to the paper industry or any paper corporation to fell is not a solution. It would be like asking Dracula to take charge of a blood bank. Putting faith in one's own people to conserve their forest is necessary. The north-east re-establishes this faith.

SECTION 3

Political Economy of Hindustan Paper Corporation (HPC)

The first question that needs to be asked about the controversial HPC is: why did government set up a PPI unit in the public sector; and the second, an even more intriguing question, why in a turbulent underdeveloped region?

Why HPC

It is important to remember that the planners of new independent India wanted not just industrial development but rounded development. Production of writing and printing paper, particularly that which would reach the mass of uneducated, as well as newsprint was a step in that direction.

Paper Situation since the 1950s and the Birth of HPC

At the commencement of the first Five Year Plan (FYP I), the annual rated capacity of the paper industry in India was 1,36,000 tonnes. Production remained more or less stationary during 1951-2 and 1952-3. However, during the next two years, there was a marked increase in production of paper, enabling supply to match demand.

While the tempo of industrial production as a whole had slowed down in 1958, that of producer and consumer goods industries continued to be relatively high. Noteworthy progress was recorded by industries which included paper and paperboard industry (FYP II—*Progress Report*, 1958-9).

The rate of growth of demand for paper during the first plan period was about 8 per cent per year. In the second plan period the demand was expected to show a progressive annual increase of 10 per cent. The actual increase in consumption as distinct from demand, averaged about 9.7 per cent per annum in the first four years of the second plan. This somewhat lower consumption as compared with demand estimates was explained by the planners by the fact that restrictions were imposed on the import of paper and paperboard from 1957 onwards. The supplies of paper and paperboard were therefore not sufficient to meet the full demand in the country. Thus, during this period there was a scarcity of paper, availability being admittedly short of demand.

The same second plan document, while dealing with the issue of paper prices, gives further information about the situation of scarcity of paper brought about by the introduction of a restrictive import policy. The government received complaints about the unduly high prices charged by wholesale merchants and retail dealers of paper. In order to safeguard the interest of the consumers, the Government of India, after discussions with paper manufacturers, decided to refer the question of fair prices for paper to the Tariff Commission, which later submitted its recommendations. The Government then permitted a revised price list to come into force in 1960 (FYP III—'Programmes of Industrial Development' (*Review of the Progress of the Second Plan*).

During the third plan, the expansion programme of the paper industry broadly conformed to the aim of achieving self-sufficiency in respect of writing and printing paper. An important development which deserves mention was the licensing of some small units mostly for the manufacture of cheap writing and printing paper for which there was a considerable demand.

The establishment of such units was expected to result in a wider distribution of sources supplying paper, reduction in transport charges and fuller use to be made of local raw materials such as rags, waste paper, agricultural residues, etc., which the large paper mills did not care to use (FYP III—*Programmes of Industrial Development*).

With the new units the total output of paper continued to increase in 1967 but the annual rate of expansion came down drastically from 9 to 4 per cent. The large forest based units, annoyed with the price control are said to be responsible for the apparent fall in paper production. It is believed to have been sold in the black market (see Ch. 6). Though the demand for paper evenly matched with its supply at that point of time (1967), the government felt, judging from the slowing down of the rate of increase in installed capacity, that unless immediate steps were taken to attract fresh capital to the industry a shortage of paper might develop by 1970-1.

It appears that the only incentive the paper industry needed to boost its production capacity was on the price front. This view is further strengthened by the fact that since the introduction of price control in 1960, except for a revision of Rs. 60 per tonne granted in June 1962, paper prices remained unchanged. Hence there was a slowing down in production of paper by the big mills. However, when the Government of India announced decontrol on prices on all varieties of paper in May 1968, the production of paper rose, registering an increase of 9.1 per cent in 1969-70 as against 4.9 per cent in the previous year. The RBI took note of this.

However an important point to be noted is that the large paper producers took advantage of the decontrol by producing a larger quantity of highly profitable speciality papers either by renovating their machinery or by importing a specific type of machinery for the purpose. This was a period of renovation and modernization of paper mills. Such modernization wholly defeated the government's purpose. After all, government incentives were to increase production of inexpensive writing and printing paper required for itself and for the poorer sections. What, however, is not clear is why the government permitted import of paper machines which could produce only speciality paper.

Since none of the incentives to woo the paper industry seemed to work and, since it did not force the issue, the central government sought an alternative in terms of setting up a corporation of its own to especially solve the problem of production of writing and printing paper. The Hindustan Paper Corporation (HPC) was mooted in the late 1960s. In 1969 the Congress had split and the ruling Congress (Congress-I) was still not quite stable.

The HPC was finally set up in 1970 to generate additional capacities for writing and printing paper and newsprint, at a time when there was stagnation in the output of printing and writing paper. NEPA was in trouble and there was significant dependence on import of newsprint. Thus HPC was set up to produce newsprint. The government also thought of taking over sick mills and nursing them back to health which was the philosophy of that period.

Perhaps the ideas of the Government of India were laudable. Perhaps it did not conceive of HPC as a profitable industry. But why was HPC not conceived as a unit that could grow on its own to produce goods that were urgently needed on a sustainable basis? Did the Government of India imagine it could bring about development by setting up units in totally underdeveloped areas?

TABLE 9.6: PRODUCTION OF HINDUSTAN PAPER CORPORATIONS PAPER AND NEWSPRINT UNITS

Name of the mill	Capacity (tpa)	Product
Hindustan Newsprint Ltd. Kerala—Subsidiary HPC	80,000	Newsprint
Mandya National Paper Mills Karnataka—Subsidiary HPC	18,200	Writing & Printing
Nagaon Paper Mills Assam (Unit)	1,00,000	Writing, Printing & Newsprint
Cachar Paper Mills Assam (Unit)	1,00,000	Writing, Printing & Newsprint
Nagaland Paper & Pulp Co. Nagaland—Subsidiary HPC	33,000	Writing & Printing

Source: HPC, HNL, MPM and MNPM.

Incidentally, it is important to mention that India had asked FAO's assistance for identification and exploration of alternate raw material in 1973. It recognized that a new technology that would save the raw material and would reduce or even eliminate costly sulphate and even long fibre pulp was necessary. The Government of India's operating agency was the HPC. It is not clear why HPC team was replaced by team headed by Ministry of Industry by the end of 1978. UNDP-FAO provided a full-time project manager. In other words FAO was involved in the HPC project.

Initially UNDP's contribution was $1.35 million and the Government of India's Rs. 17.45 million. The project became operational in 1975. In 1977 the tripartite review committee made a request to extend the project to 31 December 1980. The UNDP's final contribution was US $ 42.06 million and the Government of India's Rs. 31.39 million.

True, there is abundant raw material but nothing else besides this. The government did think that HPC would develop new technologies and consultancy services to set up a marketing network to make paper available to all sections of the community at fair prices so that malpractices during periods of shortages could be curbed (Dagli 1979: 7).

Though the HPC was set up with lofty aims it has turned out to be exceedingly dissatisfactory. There are many skeletons in this public sector unit's cupboards. Therefore two decades after its inception, HPC has not only not achieved its objectives, but today is in dire need of being rescued. It needs massive central doles and concessions for its survival. Why?

Today with a paid-up capital of about Rs. 369 crore, HPC has invested nearly Rs. 850 crore in its projects, but the corporatiopn has little prospect of earning any returns on this massive investment. Why? The apparent answer seems to lie in the location of its two major units and the dead weight of the newsprint-producing Mandya and Tuli units. The success of Hindustan Newsprint Limited (HNL) re-establishes this point.

HPC started its careet saddled with the responsibility of nursing a sick paper mill. Mandya National Paper Mills Ltd. (MNPM) of 1962, was taken over from the private sector, the reasons for which are not clear (*Financial Express*, 7 September 1986).

There was a great hue and cry when HPC acquired MNPM; no one even pretended that the mill would be profitable. HPC took it over under pressure from the central government and politicians' promises, which were never fulfilled by the Karnataka government (*Update*, 20 February 1987).

The MNPM, with an installed capacity of about 11,000 tonnes a year, was an unique plant based on bagasse. After the takeover, HPC made persistene efforts to step up MNPM's output but the Karnataka Paper Mill sank deeper into the red. The decision to use a high percentage of bagasse in the finish, and manufacture heavy grammage paper to utilize the paper machine to its full capacity proved abortive, as basic input economics turned adverse, leading to higher cost of production.

HPC may have undertaken a Rs. 20-crore rehabilitation programme to bale out MNPM, but the basic problems remained untackled, viz., material mire, mismatching of targets and production systems, technology, tardy responses to local problems, etc. In April 1986, the MNPM had to close down due to power and raw material constraints. The mill was referred to the Board for Industrial and Financial Reconstruction (BIFR) as its full worth has been eroded (*Financial Express*, 7 September 1986 and *Update*, 7-20 February 1987).

If HPC is not doing as well as it should there are several other questions that need to be answered. To put it down to locational problems *per se*, without answering why three different locations were selected in the north-east, evades the main issue. We do not deny that the main issue is locational, but we would like to ask why has none of the perceptive private sector paper and rayon corporations failed to set up its unit in this unimaginably extensive bamboo rich state? And, why were Jagi Road in Nowgong a potentially tense area, Panchgram in Cachar a highly seismic area, and Tuli in Nagaland which does not even have basic telephone connections, specifically selected?

Parameters for Selection of Site

Were the parameters for selection of the sites and time for setting up the units by the central government correct? A comparison with the parameters set by the private sector PPI would be salutary. The emerging Indian bourgeoisie, the first three paper giants, and the forward-looking Diwan of Mysore—the founder of modern Mysore as well as of Hyderabad—timed their entry correctly, viz., when the political power of the colonial British was waning. It was a time when provincial governments were elected to power by popular vote. It is no coincidence that 1937-8 marks the formal date when five Indian paper mills were set up. Locationally, too, they chose areas which were safe. These were situated either in provinces which were under Congress power or in the native states. The Government of India's reason for setting up HPC was totally different. It was primarily a defensive reaction rather than a positive one. As a result it paid little attention to either time or place. The seventies was a period of increased turmoil in the north-east. Right from the 1950s there had been movements by the hill people of the north-east against the centre's neglect. This continued into the 1960s and was compounded by the language issue. In the 1970s the Congress was in difficulties and emergency was declared. In 1979 Assam was plunged into a state of turmoil, so much so that no elections could be held in the Brahmaputra valley in 1980. The Nellie killings near Nowgong and burning of numerous wooden bridges in that state, highlight the 'instability' in the region that lasted till the mid-1980s, the period when HPC units were being set up. These problems will have to be borne in mind to understand the repeated delays and escalating costs, when the question of setting up the factories is discussed later. Logically and finally, the units went into production in 1985 and 1987 in Nowgong and Cachar respectively.

If one goes back in time, one can notice that the Indian PPI giants could only take off after the transfer of power because of state support to the industry in the form of raw materials given away at throwaway prices, besides locating electricity grids, building a network of state highways which would run close to the paper units, to boost paper production, etc.

Attracted by the infrastructural facilities and an almost monopolistic structure of the industry which was so well supported by the state, others like Thapar and Somani jumped into the profitable paper industry in the 1950s. The southern states, not to be left behind, but fighting shy of Hindi, particularly Marwari capital, scouted around for local entrepreneurs and strengthened them with state resources sometimes successfully and other times not, to set up paper units. They also threw open their doors to international capital, as and when necessary, to build PPI units.

Locational Parameters

The entrepreneurs, both old and new, had carefully selected their locations to fulfil their basic requirements: (a) Plentiful supply of bamboo, (b) A perennial stream, (c) Power, (d) Readily available cheap casual labour, (e) A railway line not too far from the potential site to transport goods and a highway. The first batch of entrepreneurs who set up their factories particularly in the central belt as well as in Mysore had the added advantage of getting coal and lime in the vicinity, thereby establishing a variety of backward linkages to profitably produce not only paper but also bamboo and wood-based rayon

Why HPC in the North-East

The question is why did the Government of India put up its units in the wrong place, or rather, permit itself to put them up in three wrong places in the north-east? The answer has to be sought in politics, that is, in electoral politics.

In electoral politics, a bourgeois state cannot offend but must placate its electorate. No paper giant had found the north-east worthy of heavy investment despite enormous quantities of bamboo and water. HPC could therefore be located in the north-east without any offence. FAO had repeatedly recommended the north-east (FAO, 1954, 1970). Theoretically, perhaps the setting up of a core industry would placate the electorate. Indira Gandhi's forte was politics, not economics of development. In the late 1960s she had found it expedient to split the Congress party. At an all-India level she tried to win people through populist techniques like nationalization of banks and insurance companies and *Garibi Hatao* (remove poverty), programmes. In Assam, her technique was to encourage the backward classes, the 'black' tribals and the Muslims by inducting them into positions of leadership. This alienated the Asamiya middle class, the powerful mouth-piece of the Congress—and created a crisis for the party. The Congress President, Devkant Barua, a native of Nowgong supported Mrs. Gandhi, eulogizing her. But Assam was not united.

Upper Assam's contradiction with Cachar is of a different geo-historical nature and no ruling party can allow this contradiction to sharpen as it is a sensitive frontier area. For this same reason the Government of India chose to locate a unit each in two areas, the Brahmaputra and Barak valleys. The specific selected spots are much less significant, they are linked to local political leaders and are valuable vote-banks in a turbulent region.

There were other contradictions between the two valleys. The post-colonial Barak valley was not only the more backward of the two, it was also a politically sensitive narrow corridor not only with Tripura and the restive hill states but also with an angry Bangladesh in the

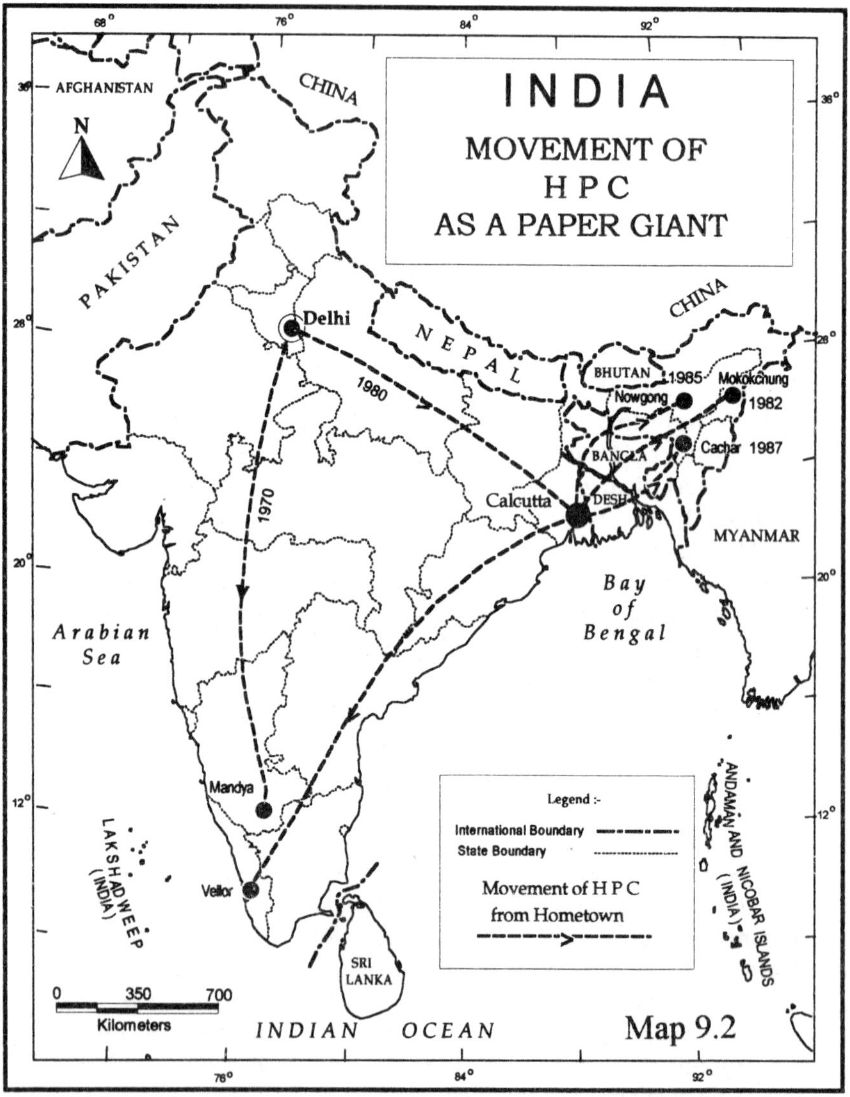
INDIA
MOVEMENT OF
H P C
AS A PAPER GIANT
AFGHANISTAN
CHINA
PAKISTAN
NEPAL
BHUTAN
CHINA
BANGLA
DESH
MYANMAR
Delhi
1980
1970
Calcutta
Nowgong
1985
Mokokchung
1982
Cachar 1987
Mandya
Vellor
Arabian Sea
Bay of Bengal
LAKSHADWEEP (INDIA)
ANDAMAN AND NICOBAR ISLANDS (INDIA)
SRI LANKA
INDIAN OCEAN
Legend :-
International Boundary
State Boundary
Movement of H P C from Hometown
0 350 700
Kilometers
N
68°
76°
84°
92°
36°
28°
20°
12°
Map 9.2

background, and could not be alienated. So when Moinul Haq Chowdhury, a powerful Congress MP, demanded a unit to be set up in Cachar the same year as in Nowgong, it was readily agreed to. The brain behind this move is believed to be Altaf Hussain, a protégé of Indira Gandhi.

The contradictions between the two regions are also historical. Devkant Barua had to be rewarded for his support. Thus it was decided to set up the first HPC unit in Jagi Road in Nowgong district in 1971.

Setting up a paper factory in the Barak valley was not based on the older Indian planners' notion of development of backward regions. If it was, then it is the Barak valley more than the Brahmaputra which needed to be developed, even if the infrastructure had to be laid. In other words, the basic rules of economic development must be followed before high investment is made for sophisticated paper mills to be set up. Setting up of the two HPC units almost simultaneously in the valleys was an act of political exigency that Congress set in motion, more than two decades ago, and which the Janata in its brief period of power continued. The Congress in turn took it on and was then keen to get rid of the mills.

The government's folly in setting up the two units in Assam was preceded by yet another blunder, namely, taking on an even more unmanageable Nagaland State Sector paper unit. Why did the Government of India weigh down its new HPC with another unmanageable sick unit? Again, the answer is to be sought in the history of Nagaland.

Hindustan Paper Corporation—Controversy

The country's largest public sector paper manufacturer, the Hindustan Paper Corporation (HPC), is caught in a bind that may ultimately threaten its very existence. Two of the units Mandya and Nagaland units that it took over have had to be closed down and two others that it set up in Assam were offered for privatization in 1993. Hindustan Newsprint Limited (HNL) continues to do well. Although three of the HPC units are operating at above average levels of efficiency, the corporation as a whole finds itself drifting and facing a not too bright future at a time when the paper industry needs every addition to capacity and production that it can muster.

Due to a multitude of reasons, along with the entire paper industry, capacity utilization has been low for the HPC. In 1992-3, HPC produced 2.3 lakh tonnes of writing and printing paper and newsprint as against an installed capacity of 3.3 lakh tonnes. Not only does this represent a 69 per cent capacity utilization, but is only marginally lower than the production target of 2.4 lakh tonnes set for the year.

HPC began in 1970, by taking over the sick Mandya National Paper Mills. Another failure was the takeover of Nagaland Pulp and Paper Mill (NPPC), and also the wrong sites chosen for its north-east units, which has been discussed in the earlier section. The one argument in favour of locating mills in the north-east was the abundance of bamboo and pine (*Pinus Khasya*) forests and perennial rivers. FAO (1955: 91-2) had suggested locating the mill at Guwahati. This strategy has completely backfired. For both NPPC as well as the Assam projects, apart from bamboo, procurement of other raw material is a big problem due to poor infrastructure and lack of feeder roads.

A special article on HPC points out that it was easy for the top management to foresee that these projects would lose money heavily. This was because operational constraints like power shortage, poor infrastructure, difficulty in attracting skilled workers to the area, would all prevent them from getting over the break-even hump (*Update*, 1987).

Worse still, even if HPC managed to procure enough raw materials and keep its mill busy, it was known that it could not sell paper locally on commercial terms, the local paper consumption being only 25,000 tonnes per annum in 1987 (*Update*, 1987). That is exactly what is happening although as much as 92 per cent of HPC's writing and printing paper production capacity (2,38,000 tonnes) is in the north-east. HPC's paper, therefore, has to travel all over India in order to be sold, which obviously pushes up the price. Unfortunately, unlike many other public sector companies, HPC is not a monopolist; it has to compete with large and small private sector mills for sale of its printing and writing paper, and with joint sector units for the sale of newsprint. It is losing out to both in the market place. For instance HPC finds it hard to sell paper to Sivakasi in Tamil Nadu, because its freight cost is twelve times more than that of the southern mills. The small paper mills were also encouraged by the Government of India when the powerful paper corporation defied and failed to produce the necessary quota of levy paper. By simultaneously putting up HPC units too, the Government of India has created contradictions between the two. All the locational problems 'could have been easily visualised then', says a senior manager who has put in more than ten years in HPC. But far from opposing the location choices, HPC's top management is reported to have joined hands with the Industry Ministry and the Public Investment Board in these foolish investment decisions, by misguiding them about the possible rates of return on these projects (*Update*, 1987).

Of course, New Delhi played a big part in HPC's fortunes by faulty location choices based on political considerations, and later, when it was compelled to set up small paper mills to overcome the paper famine created by the big mills. The government rained liberal concessions on the small paper mills, thus helping them to outprice HPC products from a near-stagnant and competitive market in the 1980s.

HPC was expected to be the government's answer to the paper shortage of the 1960s which had hit the educational sector hard, even as the private mills and the paper traders profited. By setting up large writing and printing paper capacities, it was hoped that imports could be curbed and market malpractices checked by filling supply gaps and preventing shortages. Artificial shortages were permanent then, because even as the demand rose, the private sector mills shied away from investing in the capital intensive and long gestation-prone paper industry. HPC was dealing with imports and sale of writing and printing paper till its two Assam and Kerala units went into production.

In 1978 when the Nowgong and Cachar projects were germinating, the government decided to encourage the small-scale paper mills by allowing them to import pulp because pulp-making is capital and energy intensive, doling out huge concessions on excise (25 per cent) and sales tax (8 per cent), and even after keeping them out of the Paper Control Order, by which the big paper mills were under an obligation to sell a third of their output to the educational sector at concessional rates.

Extremely unstable conditions in the north-east during the 1970s, which stretched on till the mid-1980s have already been indicated. These measures added up to a cripplmg blow for HPC because they reversed the market scenario. In 1987, when the Cachar Paper Mills, HPC's units in the north-east, started commercial production, there was a glut in the market. The Paper Control Order had been lifted and all the big mills were left with unsold stocks. In this sense, all the three original aims of setting up HPC in general and the Assam projects in particular can be said to have become irrelevant at that moment. The nation is now saddled with about Rs. 800 crore of core-sector investment that will only suck in more money from the exchequer to stay alive; unless some drastic action is taken to improve the infrastructure and permit the units to function as business units without interference and a pro-private sector bias. HPC has failed to lobby hard against the problems New Delhi heaped on it. Some senior managers reported that the top management showed little interest in protecting the Corporation's interests and that decision-making had largely been aimed at promoting personal positions. But it is also very true that the ministries allow the managers of its public sector units little space to function efficiently. Under the existing relationship, the industry ministry claims that it allows operational freedom to HPC while retaining overall control based on periodic results. This arrangement however, is said to have crumbled totally because neither the industry ministry nor HPC has done its part of the job properly (*Update*, 1987).

Marketing and Operations of HPC

Although HPC was mainly concerned with marketing in the initial years, there is no information available before the 1980s. In 1980, with no products of its own to sell and therefore no generation of internal resources, HPC was fully dependent on government loans and subsidies. The only redeeming feature, however, is the marketing wing, which earned some money by selling imported paper (*Business Standard*, 14 May 1980).

During 1981, HPC reduced the prices of its imported paper by Rs. 300 a tonne from its earlier price of Rs. 8,500 a tonne for the 3,000 tonnes of stock for sale. This eased the price situation in the economy. An HPC spokesperson said that paper was the only essential commodity where imports had not only checked the price but had actually helped to bring it down. The imports, he said, had effectively brought down not only the rise in price by about Rs. 1,000 a tonne, but the same amount of premium that was being charged in the trade channel. Between 1978 and 1979, when the shortage 'occurred', the price of indigenous paper went up by Rs. 1,000 a tonne and another Rs. 1,000 was being charged from the trade as 'on money'. The reason behind the scarcity created in paper was because of the monopoly, eight houses controlling 80 per cent of production. The government had no control over their production except for 30 per cent levy paper. It was alleged that a shortage could be created by holding back even one month's inventory either by the mills or by the trade as seen before (*Business Standard*, 18 August 1981).

In 1981-2 HPC marketed a lower volume of imported paper, totalling 20,634 tonnes compared with 31,199 tonnes in the previous year. This was due to improvement in the availability of paper from domestic sources. The paper manufacturers in this period made a number of pleas to the Government of India to withdraw production and price control order so that they could make good a part of their loss in the prevailing depressed market in the late 1970s. It is pertinent to note that because of the timely intervention of the HPC as advised by the government, the market price of white and printing paper could be brought down to a reasonable level from a prohibitive one in 1979-80 (*Financial Express*, 30 June 1982). Again, we have no information of what happened between 1982 and 1985.

HPC in its annual report for the year 1985-6 stated that the overall market demand in respect of all varieties of paper was extremely sluggish which affected the speedy off-take of paper from its units. The open market prices also continued to show a declining trend. Against this background HPC, with concerted sales efforts, marketed 16,454 mt of writing, printing and wrapping paper from its two subsidiary mills, viz., MNPM and NPPC (Nagaland), as also from Nowgong Paper Mills (Nowgong). In addition, it also sold 79,074 mt of newsprint from HNL, saving foreign exchange of Rs. 65.73 crore. However the net loss of HPC was Rs. 22.47 lakh in 1985-6.

The mill-wise sales are given below for the year 1985-6:

	Quantity (mt)	Value (Rs. in lakhs)
MNPM	6,352	680.48
NPPC	6,131	627.58
NPP	3,971	421.14
HNL	79,074	6,573.76

Source: Annual Report, 1985-6.

The sales position of HPC for the same year as given in an article on HPC, says that paper worth Rs. 10 crore was lying unsold. This is the period when the big paper mills had created a glut. But the reasons given for this accumulation of stocks, according to market circles, was the failure of HPC to compete either in price or in quality. It is also true that HPC was unable to market its products at competitive prices as its cost of production was too high due to costly pulp import. There was also a persistent complaint among the consumers that the terms and conditions for marketing HPC products were not very satisfactory. The corporation was able to make some profits from sale of imported paper in the beginning. The easygoing activities for marketing foreign products also proved to be unequal to the task of aggressive marketing which was definitely needed to sell indigenous paper (*Business Standard*, 21 January 1986).

HPC's loss-making position remained unchanged even during 1986-7. It was thought that HPC was beyond redemption even when some of its projects in the north-east were fully operational. Cachar Paper Mills, NPM and NPPC continuously incurred losses while the glut in the paper market made it difficult for HPC to sell whatever little it produced except the newsprint made by its subsidiary HNL, the lone profit-earner in HPC's fold. The products of HPC's Assam and Nagaland mills could not compete in prices and quality in the market place. The corporation was, therefore beset by a rising inventory problem straining its resources even further. According to HPC officials, the fiscal scales were tilted in favour of smaller private sector units till the mid-1980s. The bigger integrated mills had to pay excise at Rs. 2,200 a tonne, while it was only about Rs. 275 for smaller units. Small mills accounted for 40 per cent of the industry's installed capacity. Following liberalization, an investment boom was seen in the 1970s in the small-scale paper mills sector. These units were set up at smaller capital cost, based on imported pulp, wastepaper and straw. They made no investment in pulping and soda recovery plants, etc. Large scale imports of second hand paper machines with a capacity of 30 tpd or even less but with reserve capacity, added to their profitability.

After allowing the public sector a pivotal role in the development of the paper industry, the government's decision to simultaneously encourage small entrepreneurs to set up paper units created contradictions for its own units in the north-east. In other words, government support to small-scale paper units proved to be a double edged weapon. Explaining the problems faced by HPC, its managing director Maj.-Gen. V.P. Yadav said that the product mix of Assam-Nagaland mills caused a severe competition in printing cultural grades paper market. Besides, under the levy system, producers were expected to subsidize consumers. According to him a control on the industry would serve the purpose instead of a levy system. Yadav seems to have forgotten that the large paper mills were fighting the government against the levy system, and it is precisely the reason why the Government of India chose to set up HPC, and also to encourage the small entrepreneurs.

True, the industry's installed capacity of 25 million tonnes a year (1986-7) was far ahead of the domestic market demand estimated at 15 million tonnes, while indigenous production was only about 16.25 million tonnes, resulting in a large mill-made glut. HPC had a capacity of 2.33 lakh tonnes a year in its north-east mills. 'If they produced even 70 per cent of their capacity, where is the market for their produce? Maj.-Gen. Yadav asked. He suggested that the union government should consider steps like freight equalization for paper produced in underdeveloped regions like the north-east, a substantial excise relief and corresponding sales tax relief by the state governments, together with improved operational efficiency of the mills with a more flexible product mix. All these factors would certainly greatly help HPC (Vinayak, *Financial Express*, 7 September 1986).

Another reason for HPC's poor performance was the frequent changes in bureaucracy at the ministry level and top management of HPC, depriving the corporation of the benefit of 'continuity' in management. Since its inception, few had held the top position for more than two or three years, except M.L. Zutshi, a protégé of Karam Chand Thapar and Co. which runs the profitable Ballarpur Paper unit. The corporation had a middle level management which apparently lacked commitment. The same was the case with its marketing department which did not have a general manager for as long as seven months, after the last incumbent left in August 1986. HPC was said to be without a head. Under the circumstances, in 1987 HPC had unsold stocks of nearly 12,000 tonnes while its losses continued to mount. Accumulated losses were estimated at over Rs. 130 crore (*Financial Express*, 8 March 1987).

HPC's working results changed drastically in the year 1988-9. Despite suffering a continuous drag on its exchequer in the shape of its loss-making subsidiary NPPC, it was showing signs of recovery. The recovery path however, was laid by HNL, which achieved 98.3 per cent capacity utilization and the surplus of Rs. 2 crore, it was considered a jewel in HPC's crown. The year 1988-9 closed with a higher operating profit of Rs. 22.08 crore against the previous year's Rs. 11.29 crore before depreciation and interest, producing 1.86 lakh tonnes of newsprint, writing and printing paper, and posting a growth of 8.14 per cent. HPC was therefore determined not to seek any cash assistance from the government during 1989-90 because of a sharp improvement in its performance. However, the continued poor show of its subsidiary—NPPC, was causing the management a great deal of worry (*Financial Express*, 9 November 1989 and *Business Standard*, 5 December 1989).

In 1989-90 HPC achieved an operating profit of Rs. 71.77 crore. Both Cachar and Mandya National Paper Mill which were in the red, recorded operating profit during the year. HPC's installed capacity was 2.51 lakh tonnes of paper and 80,000 lakh tonnes of newsprint. The capacity utilization of the corporation was 73 per cent which compared favourably with the national average of 63 per cent. The product mix was also altered so as to have higher value-added products (*Financial Express*, 3 May 1990). However, another article on HPC in the *Economic Times* gives quite a different figure of HPC profits—Rs. 71.77 crore during 1989-90. The reasons for this manifold increase in profits was the increased volume of operations.

Because of the success it achieved during the year (1989-90) HPC was being 'reasonably optimistic' about its immediate future after being buffeted by misfortune, mismanagement and mishandling for an incredible 19 years, remarks an *Economic Times* article on HPC's success. HPC was in a commanding position in the Indian paper industry, being the largest producer of paper (including newsprint) and for the first time, since its inception in 1970 was seeing the 'light at the end of the tunnel' as one of its senior executives remarked. The other remarkable aspect of the year's performance was HPC's decision to go without government's non-plan support for the first time and also, repayment of Rs. 7.6 crore towards a loan from a much earlier payment. Such confidence by itself is rare in public sector corporations which, like many other corporations, had been saddled with politically motivated decisions leading to impossible techno-commercial situations. (*Financial Express*, 13 November 1990 and *Economic Times*, 3 May 1990).

In 1990-1 the operations of the company experienced mixed fortunes. HNL continued to maintain its encouraging track record of performance. Cachar also witnessed an upward trend in output and profitability. Production at the Nowgong Paper Mills, however, showed a decline due to the disturbed socio-political scenario in the region. The capacity utilization for the year was 77 per cent, marginally higher than the previous year's level of 74 per cent. The operations of these four mills resulted in an aggregate net operating profit of Rs. 78.25 crore, an increase of 8.86 per cent over the previous year's Rs. 71.77 crore (*Annual Report*, 1990-1).

Early in 1992, HPC was burdened with at least 25,000 tonnes of cream wove paper valued at not less than Rs. 50 crore. It could not afford to ignore the stock pile-up as this public sector company was facing a severe fund crunch and had Rs. 540 crore of accumulated loss. In an effort to boost its sales, the HPC Chairperson and Managing Director, B.T. Sridharan, held a meeting with two accredited dealers. The dealers wanted special discount systems at the flat rate of Rs. 800 per tonne in addition to the normal trade discount of Rs. 900 per tonne. HPC came out instead with a special trade discount rate on the slab basis—Rs. 800 per tonne if the dealer lifted 100 tonnes in one month, Rs. 1,000, if the lifting was for two consecutive months, and Rs. 1,100 if it was for three consecutive months. But the trader was not happy with this slab system as such benefit would accrue only to big traders, while small and medium traders were hesitant to deal with HPC. The dealers believed the location of HPC's mills to be yet another negative feature. HPC's cream wove paper is sold at an average rate of Rs. 20,000 per tonne, inclusive of duties and trade discounts. This price was fixed in September 1991, after an upward revision of rates. Though HPC has not rolled back prices unlike some other big paper producers in the country, the incentive offered to the trade itself was an indication of its nervousness about the market position (*Business Standard*, 30 March 1992).

At the close of 1992, the centre wanted, among other alternatives, to offload the whole of HPC to any suitable private sector entrepreneur (*Business and Political Observer*, 1 January 1992). A couple of private sector entrepreneurs belonging to the Thapars and Birlas visited NPM and Cachar Mills in Assam under the initiative of the Union Industry Ministry, but the feedback was negative. However the HPC MD is believed to have said to the HPC Officers Association that it was for the union government to sell off HPC, and that he was unwilling to take any initiative towards this. HPC in an official press release also made it clear: 'We would emphatically like to point out that the Government has not taken any such decisions in the matter. In fact, the Government has under consideration a number of options for further consolidation of HPC's mills in the north-east. This includes, inter-alia, a capital restructuring programme, technology upgradation and even disinvestment.' The company confirmed that the SBI Capital

Markets Limited was undertaking a scrutiny of various alternatives which include even privatization. HPC's corporate management is pursuing the centre with its proposal for financial restructuring and technical upgradation of the company. Cachar has already achieved a 90 per cent capacity utilization while Nagaon has crossed 70 per cent. The HPC corporate forecast is that they will fulfil the target and achieve a marked improvement in techno-economics (*Business and Political Observer*, 1 December 1992).

Much of HPC's future would depend on the decision regarding NPPC, MNPM and the two units in Assam. Far too much uncertainty lies in store for the corporation which is also facing a declining profitability. Although there is a steady rise in the turnover of HPC—at a pace of 10 per cent every year since 1991-2, profitability is eroded by the fact that as the only net profit-making outfit of HPC, HNL has started to pay income tax resulting in a dip in net profits. The position regarding the government's newsprint policy does not make the situation a particularly bright one for HPC. To put it in the words of HPC's CMD, Sridharan, 'If Cachar and Nowgong mills are privatized and Mandya is closed down and NPPC revival proposal fails to get underway—HPC will cease to exist' (*Update*, 1993).

It is ironical that the government established HPC with fanfare in order to discipline the errant big paper industry which indulged in malpractices for windfall gains in the 1970s but two decades later, the government's own policy changes, together with the difficulties that HPC is saddled with, has only turned the market tide in favour of the private sector (*Update*, 1993).

SECTION 3 (A)

Nowgong-Jagi Road and Cachar-Panchgram Paper Mills—HPC's Units in Assam

HPC was set up in 1970, while seven years later, Nowgong and Cachar districts which were well endowed with bamboo forests were selected for locating the two PPI units. Bidyut Sarkar writes that the Nowgong and Cachar mills, each costing Rs. 115 crore, were approved by Indira Gandhi's cabinet, but subsequently came under the new government's suspension of investment decisions pending a review. The fate of these projects became a matter of speculation particularly as the Congress regime had sanctioned them just before the March parliamentary poll. However, the new Janata government gave permission to HPC to go ahead in appreciation of the growing demand for paper in various forms. It also showed confidence in the emerging potential of the first public sector attempt in the paper field which was expected to become a dominant force in the industry (*Economic Times*, 25 May 1977). The projects were approved in March 1977, with an annual production capacity of 1,00,000 tonnes each of writing and printing papers.

The Public Investment Board which cleared these two public sector paper manufacturing mills, six months later suggested to the union cabinet that it accord priority to the Nowgong project during the final exercises on the investment decision. It was understood to have suggested to the cabinet that a gap be maintained in the construction activity between the Nowgong and Cachar projects (*Business Standard*, 18 July 1977). Hazarding two guesses for such a decision—first, there was still an uncertainty regarding 36 *bighas* of land at the proposed mill site at Cachar which had not been vacated by the Border Security Force (BSF) and secondly, the engineering layout of the site was not completed. There were no hurdles for Nowgong (*Business Standard*, 1977) and Rs. 12.50 lakh were sanctioned for site levelling and grading work at Jagi Road for the Nowgong Paper Mills. A Rs. 78 crore national tender for machinery was floated from New Delhi. Also a budget provision of Rs. 10 crore for this project was earmarked for the fiscal year 1977-8.

The total requirement of land for the Nowgong project was an estimated 600 acres for which the acquisition proceedings were launched by the state government in 1979, out of which about 175 acres of land of the Sonaikuchi reserve forest were assigned to HPC to set up its plant. In the case of the Cachar project, nearly 760 acres of land which belonged to a private tea estate were purchased under a package deal through the state government at a total cost of Rs. 4.35 lakh. The detailed project reports proposed that the plants would not only have their own matching pulping sections but also captive caustic soda and chlorine dioxide plants (*Financial Express*, 9 November 1979).

The other major contributory factors for substantial time as well as cost overruns for the Cachar unit were the peculiar geography of the project area and the complicated land-acquiring procedures. According to the revised estimates, the one lakh tonne paper mill was scheduled to be commissioned by October 1984, at a cost of

Rs. 22 crore. The land problem was mainly concerned with the homestead areas. The state government delayed in finalizing the issue of eviction of people from their homesteads. The quantum of compensation had to be jointly decided by the state government and the HPC. The evicted people were most likely to demand permanent jobs. The approach road could not be constructed as it cut across the homestead lands which were then not in the possession of the project authorities. Because of this problem, further orders for plant and equipment could not be placed as delivery orders for those items were closely linked with civil and mechanical erection. These unavoidable time overruns were estimated to raise the project cost to touch a Rs. 250 crore mark (*Financial Express*, 4 July 1982).

The communication deficiencies in Assam continued to bedevil the project. In 1984, most of the Rs. 30 crore worth of plant and machinery had already arrived. According to an HPC official, some equipment had to be hauled over to the site in a complete knocked-down condition as the rickety Badarpur Lumding metre gauge rail track and landslide prone, serpentine Jowai–Badarpur National Highway, proved inadequate to carry them as single units. Some more plant material was also piled up at Jagi Road as it could not be carried to the site because of the weak bridges straddling the highway. The HPC demanded augmentation of the wagon carrying capacity in the metre gauge as the corporation had to import raw materials from outside and transport the finished products.

For both units, Utkal Machinery (in collaboration with Voith of West Germany), and Jessop of Calcutta (in collaboration with Belliot of the USA), manufactured the paper-making machinery and the 85 metre high boiler. BHEL fabricated the turbines. Of the power requirement of 45 mw each, 30 mw was to be generated by the corporation by two turbo-generator sets. An unstarred question no. 2,229 to the Minister of Industry and Commerce states the value of machinery was Rs. 157.46 crore (*Business Standard*, 19 August 1982 and an unstarred question no. 2,229 to Minister of Industry and Commerce Affairs, dated 18 August 1985). However, during our field visit one of the Nowgong paper mills senior managers stated that till recently the boilers and turbines have never worked properly.

While the Cachar project was still struggling with the construction of its mill, the Nowgong paper mills started commercial production on 1 October 1985, sadly enough, with only 50 per cent of its installed capacity of 50,000 tonnes. Its capital outlay was Rs. 400 crore. The commercial production of writing and printing paper with one machine was 6,388 mt while the second machine was still to be commissioned. HPC's annual report (1985-6) describes persisting teething troubles, non-synchronization of the operation of the turbo-generator sets and non-availability of adequate power from the state grid as reasons for low production. HPC therefore suffered a net loss of Rs. 22.47 crore in the first six months of its commercial production. The break-up of the loss works out to Rs. 7.12 crore. Out of the total production of 6388 mts, the sale was only 3,971 tonnes (*Annual Report*, 1985-6).

During 1987-8, the Nowgong mill produced 0.65 lakh mt of paper. The increase in production from 0.28 lakh tonnes in 1986-7 was attributed to a full year's operation using the second paper machine. During the year the mill earned an operating profit of Rs. 265 lakh.

Poor off-take, with one lakh tonnes in excess of demand, coupled with higher input costs and increasing interest burdens resulted in the operating profit of NPM as well as Cachar in 1991-2. At the moment of writing, as 50 per cent of the net worth of the units has been eroded, they are under reference to the Board of Industrial Finance and Restructuring (BIFR).

PTI Corporate Trends, 10 June 1993, reports that HPC has started a programme of modifying the units to produce high value items for which an investment of Rs. 10 to 12 crore has been earmarked. In November 1993, the modified machines of both NPM and Cachar had already gone into production (*PTI Corporate Trends*, 10 June 1993) but the net loss after taking into account depreciation and interest was Rs. 11 crore. Boilers and turbo-generator sets continued to cause problems.

Commercial production of the Cachar paper mills began in April 1988, seven years behind the targeted date. The mill site covered a total area of 1,073 acres and the final cost of the project was Rs. 356 crore, three times its earlier estimate cost of over Rs. 115 crore in 1977. Unfortunately thereafter, the mill suffered a breakdown in its continuous digester and was later affected by the floods in Assam.

The Nowgong mill which had earned an operating profit of Rs. 4.66 crore however, closed in 1988-9 with a net loss of Rs. 38.12 crore after providing for payment of interest and depreciation. The installed capacity of the project (1,00,000 tpa) could be utilized only in a restricted way due to breakdowns and unreliable performance of the coal-fired boilers and their auxiliaries. This resulted in lower production of writing and printing papers which fell to 58,000 tonnes as against 64,000 tonnes in 1987-8 (*Financial Express*, 9 November 1989). In 1989-90 NPM produced 70,232 mt of paper and Cachar's production stood at 47,160 mt (*Annual Report*, 1989-90). During 1990-1, NPM's production went down to 66,125 mt, adversely affecting its profitability. The net loss suffered was as high as Rs. 21.91 crore much higher than the previous year. But the production trend of Cachar was just the opposite. It registered an increase in production to 57,623 mt from the 47,160 mt of the previous year. Correspondingly, there was a low net loss of Rs. 38.38 crore as against Rs. 46.88 crore in the previous year (*Annual Report*, 1990-1).

Production and sales figures of Cachar are given in Figure 9.2. Data from NPM was not made available.

Raw Material Requirement of CPM and NPM

A unique feature of the projects in Assam is that they were based solely on bamboo as raw material. On the basis of a pre-investment survey of forest resources it had been established that there was adequate raw material available on a sustained basis. HPC entered into an agreement with North Cachar Hills District Council and the Government of Assam on 20 July 1977, for a period of 30 years from 15 September 1979, in the case of NPM, and 15 September 1980 in the case of Cachar. The Government of Assam and the Council says the agreement 'shall make available to the corporation such forest areas as would yield annually, the required quantity of approximately 2.5 lakh tonnes of air-dry bamboos (at 10 per cent moisture content) for each of the two projects of various varieties, suitable for the manufacture of pulp and various varieties of papers [herein after referred to as 'Bamboo' (Agreement copy dated 20 July 1977)].

The forest areas reserved for HPC for supply of bamboo are as follows:

In 1990, the Times of India's correspondent, wrote that the only advantage from the location of the mills in Assam had been the availability of bamboo and wood. But even this advantage was going to be lost because of disputes between the mill authorities on the one hand, and the Assam government and the local district hill council on

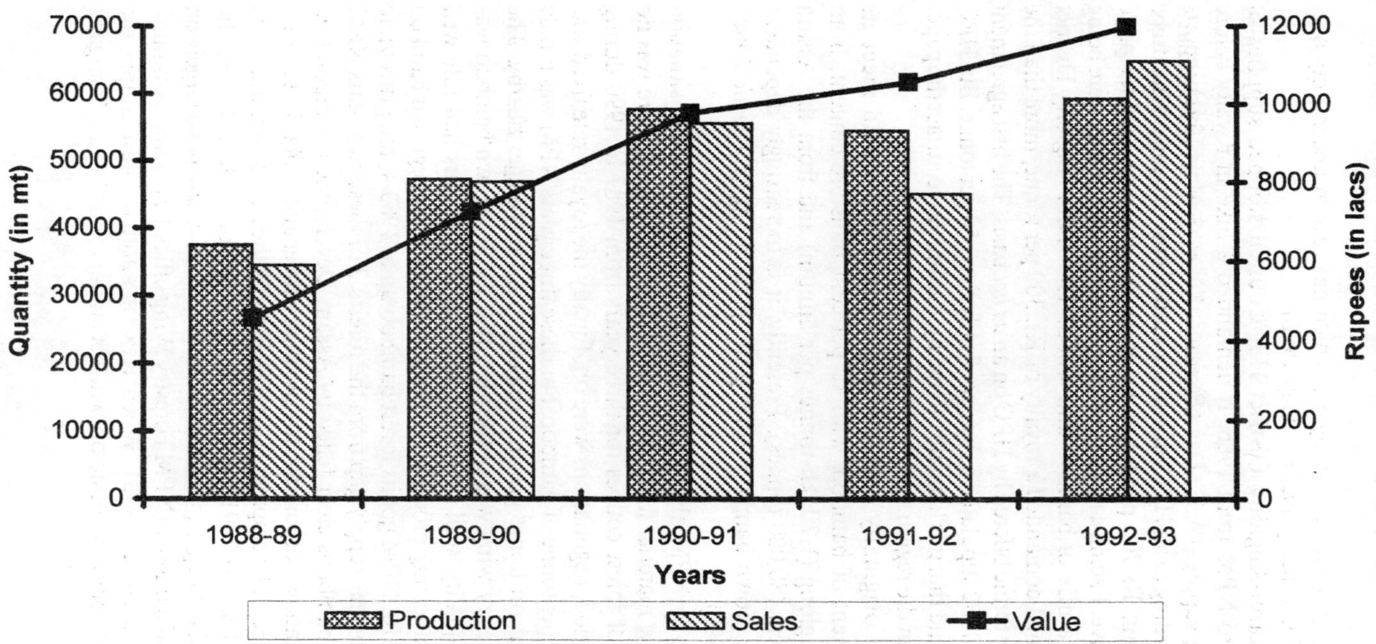

Source: Cachar Paper Mills, November 1993

Figure 9.2: Production and Sale of Paper by Cachar Paper Mills (Panchgram) from 1988-9 to 1992-3

the other, over the rate of royalty which has caused uncertainty over supply (*The Times of India*, 7 May 1990).

The Karbi-Anglong District Hill Council stopped supplies of bamboo to NPM from 1986. It demanded a hike in royalty before resuming supply—A.K. Das (*Economic Times*, 3 May 1991) reports that according to an HPC official the landed cost of bamboo had to be kept within Rs.700 per tonne to make the finished product price competitive. Facing a raw material crunch, the HPC plants were being fed with a mix of bamboo and wood. The Karbi-Anglong District Hill Council demanded a royalty of Rs. 300 per tonne, more than the Rs. 20 per tonne offered by HPC on an *ad hoc* basis. The first agreement signed in 1972 stipulated a royalty of only Rs. 2.50 per tonne. But HPC officials said that this low rate offered by Assam was to attract the paper industry in the region.

It had originally been envisaged that NPM would meet its require-ment of bamboo, with 10 per cent moisture content, from Karbi-Anglong (1.7 lakh tonnes per annum) and from the Assam government (80,000 tonnes). Presently, it is increasingly dependent on home grown bamboo to keep the plant going (*Economic Times*, 3 May 1990).

Currently NPM also seems to be in an uncomfortable situation on the raw material front. Because of the royalty issue there was no working of forest coupes for three years from 1987 to 1990 during which period, negotiations were on. Finally the royalty rate fixed was Rs. 62.50 per tonne of bamboo. For these three years NPM was totally dependent on home grown bamboo from the villages nearby. The NPM forestry wing was of the view that locally grown bamboo was of good quality, well managed, and that the villagers had followed proper harvesting procedures. The rate paid for homegrown bamboo was Rs. 610. NPM got its forest bamboo up to 1983-4, after this there was a two-year supply gap from the forests. For these two years NPM used the bamboos clear felled by Assam Development Corporation and Plantation Corporation to raise their rubber and coffee. But Assam has no Forest Development Corporation said the PCCF, Assam. From October 1988 to 1991, softwoods, i.e. firewood collected from the forest by the villagers, was used. In future, NPM will depend largely on homestead bamboos (NPM Forestry Wing official, personal communication, November 1993). When HPC is allotted forest coupes the working is done by contractors who get labour from Golpad, Marigaon and Nowgong. The percentage of tribal labourers is generally low. Contractors undertake felling, loading, transportation, unloading and stacking at the mill premises, and are paid on per-tonne basis (personal communication, Nowgong Paper Mills Forestry Wing, November 1993). Details on bamboo as run in Assam have been discussed in the earlier section on the Industrial Management of Forests and Local Management of bamboos in Section 1.

(Nowgong mill has not provided us data they promised for analysis.)

Other Inputs

The sheer magnitude of the locational disadvantage faced by HPC units in Assam could be gauged from the fact that out of total 10 tonnes of input required to make one tonne paper, as much as 5 tonnes have to be brought from distant regions. These inputs include coal, lime, salt, soap, stone, salt cake and rosin brought from as far-off places as Kanti in Madhya Pradesh, Gujarat, Rajasthan, Kashmir, Himachal Pradesh, Bihar and West Bengal. Above all, the quality of all the inputs is low and has resulted in ageing and corrosion of machinery. Bamboo and other woods are the only raw materials locally available (*Field Work*, November 1994).

For the power requirements of 30 mw the Assam government is committed to supply 10 mw to Cachar and NPM, the rest is met by in-plant power generation (*Financial Express*, 29 November 1979) However, the power position of these units in 1990 was precarious. A.K. Das writes that out of a 10 mw committed supply from the Assam Electricity Board the highest ever recorded was only 5 mw, which forced the caustic chlorine plant to be run at much below the rated capacity, and also wrecked its vital equipment (*Economic Times*, 4 May 1990).

PTI corporate trends give the 1993 power position of the two HPC units in Assam. The Assam State Electricity Board announced a stiff increase in tariff from Rs. 1.75 per unit to Rs. 2.45 per unit, the costliest in the country. This placed an additional burden of Rs. 600 per tonne HPC, according to Sridharan the CMD. On a yearly basis, it would mean an additional expenditure of Rs. 4 crore on this account at these mills every year. Naturally, the HPC management has accorded top priority to energy conservation measures (*PTI Corporate Trends*, 10 June 1993).

TABLE 9.7: RAW MATERIAL PURCHASED BY CACHAR PAPER MILLS FROM OUTSIDE ITS CATCHMENT AREA

Year	Source	Type of material and quantum	
		Bamboo (mt)	Wood (mt)
1988-9	Home grown (Barak valley)	12,634	—
	Mizoram	18,868	—
	Tripura	9,487	—
	Meghalaya	409	—
1989-90	Barak valley (home grown)	288	14,385
	Mizoram	35,128	—
	Tripura	26,594	—
	Meghalaya	1,928	—
1990-1	Barak valley (home grown)	—	26,150
	Mizoram	12,976	—
	Tripura	9,087	—
	Meghalaya	—	9,463
	Manipur	1,519	—
1991-2	Barak valley (home grown)	—	10
	Mizoram	801	27,766
	Tripura	27,643	—
	Meghalaya	—	15,496
	Manipur	10,606	—
1992-3	Barak valley (home grown)	51,932	—
	Mizoram	12,719	—
	Tripura	1,574	—

Source: Cachar Paper Mills, November 1994.

Production Technology and Machinery Used by Cachar

The plant at Cachar has two major technological features: first, a continuous digester for pulping by sulphate process and secondly, a five-stage bleaching unit including a built-in stage chlorine dioxide unit. Besides manufacturing of paper the unit has also a caustic and chlorine dioxide plant which was expected to go into production during December 1993 (*Field Work*, November 1993).

The Cachar Paper Mill has an installed capacity for producing 1,00,000 mt of writing, printing paper per annum including 20,000 mt of newsprint. The paper being produced is of good quality and meets the requirement of the student community in the education section, institutions, government departments and other consumers. Modern and innovative technology has been the hallmark of the HPC.

For preparation of bamboo chips for making pulp, the mill installed six numbers of drum chippers of latest design along with screens and chip-washing units for separating dust and dirt. Necessary conveyors and flumes have been installed in the system for material handling. The design capacity of the plant is about 828 tdp chips.

A high capacity continuous digester was installed for pulping in Cachar Paper Mills to ensure uniform quality pulp with saving in energy consumption. The digester is equipped with a counter current washing system. The capacity of the plant is about 376 bone-dry metric tonnes (BDmt) per day of unbleached pulp. The system, being a closed one, is also pollution-free.

The system of pressurized screening has been adopted to save energy and also to yield uniform clear pulp. The screened pulp is washed in vacuum drum washers. The residual alkali of the washed pulp is re-used in the digester plant. Black liquor from the digester plant is sent to the recovery plant for recovery of chemicals. The capacity of the washing screening plant is 376 tpd on design basis.

The washed pulp (unbleached) is then bleached. Cachar paper mill plans to commission the 'latest' bleaching technology with chlorine dioxide to further improve the quality and brightness of bleached pulp. The bleached pulp from pulp mill is treated in a stock preparation plant for making paper. This plant is equipped with two separate streets of energy-efficient refiners and a stock preparation system for two paper machines. The refiners consist of double disc and conical types.

The paper machines are also based on the latest technologies including the 'No Draw Trinip' system and the Deculator system. The mill has two paper machines of high speed, designed to operate at 150-700 metres per minute within the gsm range of 30-120. The No. one paper machine is equipped with size press to produce high quality maplitho paper. The capacity of each machine is about 200 mtd. Both the machines are incorporated with a closed hood in the dryer section to save steam and for uniform dryness of paper.

Multiple effect evaporators are adopted, incorporating advanced design features for concentration of black liquor obtained from cooking of raw materials in the continuous digester with alkali. For recovery of cooking chemicals from concentrated black liquor from multiple effect evaporators, the mill has installed a recovery furnace and boiler for combustion of black liquor and co-generation of steam. The recovery furnace and boiler is incorporated with latest technologies like cascade evaporator and electrostatic precipitator. Chemicals containing in-fly gases are recovered in an electrostatic precipitator thereby reducing air pollution. The capacity of the recovery boiler for handling of black liquor is 675 mtd with generation of 92 tonnes of steam per hour at 60 kg/m^2 pressure which is utilised in process requirement.

In the finishing plant, paper rolls, i.e. 'Jumbo rolls' produced by the paper machines are cut to smaller size in two drum rewinders and sheets by different types of finishing equipment. Reels are packed in a roll wrapping machine with kraft paper. The average production of this unit is 25 packed reels per hour. There are five number of Simplex sheet cutters in the finishing section. These cutters are equipped with shaftless unwind stands core chucking device, scissor-lifting platforms, trim-removal fans and several other latest technologies for high-speed cutting of sheets. The cut sheets are transferred to finishers' tables where finishers sort out, dress, count and pack the cut sheet reams. The average saleable product manufactured in this section is about 300 mtd capacity. The total area of the finishing section is about 7,500 sq. km. Finished products are sent to the distribution section for despatch.

SECTION 3 (B)

Nagaland Pulp and Paper Company—HPC's Subsidiary in Nagaland

NPPC's chequered history was somewhat consolidated in 1970, when the union cabinet 'approved in principle' a proposal to set up a paper plant in Nagaland. This led to the appointment of the National Industrial Development Corporation (NIDC) as turnkey consultants. The NIDC recommended a product mix of 20,000 tpa of writing and printing paper and 13,000 tpa of kraft paper. The project was cleared by the government in 1972.

HPC entered the scene in 1974 and questioned the location of the mill at Tuli for two reasons. It proposed Naginimara as an alternative site as it had a railhead while the one nearest to Tuli is 30 km away at Amguri in Assam. Secondly, NPPC would have saved the Rs. 50 lakh it spent in Tuli on the construction of an effluent treatment plant and a pipeline system as there is a fast-flowing river close to Naginimara.

Curiously, the two central government committees, headed by

T.A. Pai and B.P. Maurya, selected Tuli against opposition from HPC even after soil-testing and initial levelling had been done at Naginimara. Why? Political pressures, says a special report on HPC (*Update*, 1987). The team's own account, given earlier in Section 1 confirms this.

The central government succumbed to the pressure of a strong lobby led by Chetan Jamir, then Nagaland Industries Minister and Shashimaren Aier, a former Industry Secretary and Development Commissioner of Nagaland, both very close to Indira Gandhi. Aier, who went on to become a wealthy businessman after retirement from the IAS, belonged to the Ao tribe and backed the Tuli site for the benefit of his own community. Narrow vested interests obviously overrode commercial considerations (*Update*, 1987).

Construction of NPPC

The state government decided to locate the factory at Tuli which is at a height of around 800 ft above sea level in Mokokchung district. It is also 21 km from the nearest railway station, Amguri in Assam. The feasibility report for setting up a 50 tpd paper mill and its progress is as follows:

1964—The National Council of Applied Economic Research in its Forest Appraisal Report first suggested the possibility of one or two pulp and paper units in Nagaland.

In 1964-5—The Nagaland State Authority undertook the first survey for raw materials and called it the 'Pre-investment Survey'. It showed a possibility of acquiring 3,00,000 tonnes of air-dry cellulosic raw material annually from a forest area covering 757 sq. km or 1,87,000 acres covered with grasses, reeds and bamboo.

April 1967—second survey completed by the Nagaland State Authority, i.e. the Directorate of the State Forest Department increased the percentage of sampling, called *Detailed Survey Report* titled 'Report on the cellulosic raw materials for a paper mill in Nagaland' submitted in November 1967. Work was done only in two blocks, Tuli and Naginimara, only on local grasses bhutang and khagra, while making available grasses, reeds and bamboo to a paper mill at Tuli.

In December 1996—the project was accepted in principle by the Plannmg Commission. A committee was set up by the Ministry of

Industrial Development Grasses suitable for paper making, were flown to Paris for exhaustive investigation.

Mid-1968—On the basis of the findings of the Paris based report a detailed project report (DPR) was prepared. It suggested utilization of only local grasses and reeds. Bamboo was totally left out of the investigation by the French experts. It had suggested the use of some long-fibred pulp if need be, along with grasses. The technology to be used was kraft-batch cooking and chlorine-cum-chlorine dioxide bleaching. This was to be a single pulping street and one paper machine to manufacture cultural papers. It suggested that the entire production be based on grasses and the mill be projected for an annual capacity of 30,000 tonnes of pulp and equal tonnage of paper. The project cost was estimated at Rs. 15.5 crore with a foreign component of Rs. 5 crore.

December 1969—prior to the adoption of DPR the Government of India constituted a four-member appraisal group of experts drawn from the paper industry along with the Development Officer concerned.

February 1970—The recommendations of the expert team were finalized at a meeting with representatives from Ministries of Finance and Industrial Development and the Government of Nagaland. D.S. Kothari one of the team members of the expert committee suggested incorporation of bamboo for pulp making with equal proportion of grasses and diversifying the production pattern by installing two medium size paper making units so as to cover the local and surrounding market in a larger way. This major deviation which meant an enhanced project cost and greater flexibility in operation and production pattern, was accepted by the appraisal group and recommended to the central government.

1970—NIDC was appointed consultant to the Nagaland Pulp and Paper Company (NPPC).

Early in 1971, the National Industrial Development Corporation recast the earlier DPR by incorporating the necessary changes on the plant side and revising the cost structure on the investment and profitability side (Kothari, August 1972).

The HPC became involved in the project in April 1974 with an equity of 87.5 per cent while that of the Nagaland government was 12.5 per cent, at which time not more than Rs. 2 or 3 crore were spent. By January 1976, the NIDC estimates were required to be revised. Actual production was still a distant dream.

With a target date of commissioning set for June 1978, the total project cost added up to Rs. 62.10 crore. Meanwhile, the cost of land and civil works had gone up from Rs. 2.91 crore to Rs. 8.21 crore, and that of plant and machinery from Rs. 14.05 crore to Rs.32.28 crore, and the general price escalation provision of Rs. 10.17 crores in January 1976. Landslides, the difficult terrain, and the general price spiral were significant factors in delaying the project as well as pushing up the costs. But mismanagement and a dash of plain politics appeared to have played major roles in delaying this project (*Economic Times*, 9 May 1977).

In deciding the location of the project, neither the HPC nor the NPPC management could do much except recommend and then follow orders that went against the recommendations. Tuli was not the best possible place to have a paper factory in Nagaland. Construction of a huge plant at the site was a Herculean task as no flat land was available anywhere and several hills had to be chopped off to carve out some limited land for locating the various units. Besides, serious difficulties such as landslides were feared, because various units of the mills were being located on different hill tops separated by valleys all around, presenting operational and financial problems. The problem of constructing a township in such a terrain for 1,600 employees and the associated public, posed a still more difficult question.

In September 1974, and again in October 1975, landslides knocked out parts of the construction in progress. Taking these and other difficulties into account, HPC suggested a new site at Naginimara where better facilities for the paper mill were available. Naginimara was in the plains. After a long exchange of correspondence and hesitation, the Union Industry Ministry told the HPC on 31 July 1975, to go ahead with the construction of a 100 tpd capacity paper mill at Tuli. At that stage not more than Rs. 3 crore had actually been invested and a shift to Naginimara would have saved over Rs. 5 crore in project costs. It would be pertinent to point at this point, that the powerful Ao tribe dominates the population of Tuli and the entire district of Mokokchung, whereas Naginimara is peopled by a different sect—the Sema Nagas. The second factor responsible for the delay appeared to be the slow job work being done by the Nagaland Projects Construction Corporation (NPCC). That some serious lack of coordination existed at one time between the civil engineers and the erection engineers was evident from a glance at the dates of the railway receipts stamped on the numerous idle machines lying at the plant site. Most of the machines arrived in late 1975, but were installed only in late 1976 (*Economic Times* , 9 May 1977). When pressed to comment on this point, some officials pointed out that NPPC's work was handled mainly by sub-contractors over whom the HPC (though it had an 87.5 per cent stake in the company), had little control. The National Projects Construction Corporation on the other hand, had other responsibilities in Nagaland. So the corporation representatives were not available at the plant site for more than 15 days a month, their local headquarters being at Amguri in Assam, 21 km away.

A third factor was that the sub-contractors employed mostly non-locals who were not on the best of terms with the local people. There were some instances of threats and counter-threats and of sub-contractors running away, after taking on the job in hand. Further, as the project was located amidst and on top of many hillocks surrounded by several valleys in the absence of flat lands, a high degree of engineering skill, and further finance were required to instal each unit. The water treatment plant has been built at a hill about 1 km away from the main plant, while the bamboo chipper plant is separated by a deep ravine from the main pulp mill unit which was being bridged by a massive steel structure over which the chip conveyors have been installed. The conveyor structure with a long stretch of 192 metres, connecting the massive bridges from the chipper to the pulp mill, was a unique sight, not known perhaps to any paper mill in the country. Similarly, the coal storage area and the crusher plant are located deep in a valley beyond the plant and the effluent treatment plant with a total cost of Rs. 8 crore has been erected a distance away (*Economic Times*, 9 May 1977).

The cost of the project was estimated to touch the figure of Rs. 80 crore in 1980 against the earlier estimates of Rs. 62 crore. The cost escalation was due to the construction of residential accommodation, annual price escalation for the equipment, increase in cost of cement, steel and other raw materials, and the increase in working capital.

In its 1978-9 *Annual Report*, NPPC stated that the erection work was being completed in major areas of the mill, i.e. paper mill, pulp mill, water treatment plant, recovery boiler, etc. However the erection schedule was affected due to the delayed delivery of indigenous equipment as some of the companies failed to supply the equipment according to the contractual delivery schedules. HPC was greatly satisfied that the central government had officially announced the construction of a railway line from Amguri to Tuli. The survey work for this purpose had already been completed and the railway authorities promised that the line would be taken up on a priority basis and was expected to be completed during the sixth plan period. This was to facilitate easy transport of raw materials and finished products to and from the project site (*Annual Report*, 1978-9). Until November 1993, the railway track had not been completed.

Where consultancy was concerned, the report said that in spite of the intervention of the centre's Ministry of Industry, the position of supply drawing from NIDC, the turnkey consultant, did not improve and the company had no other option but to engage another consultant for doing the rest of the work. Development Consultants Private Limited (DCPL) was entrusted to complete the balance work at a total cost of Rs. 20,62,500 at the risk and cost of NIDC. A seven-member committee which was probing the causes of delay in setting up NPPC in 1981 was of the view that NIDC, without any prior experience in paper, took several years to complete a part of the job. NIDC the original consultant, had struck a serious blow to the project due to its inexperience and 'casual treatment'. According to the committee, the performance of DCPL also left much to be desired (*The Times of India*, 1 September 1981).

This apart, the committee found that the site in Nagaland was not suitable for a giant paper project. In the absence of an industrial culture, lack of raw materials apart from bamboo, and failure to imbibe a public sector concept at the project management level, the committee members were apprehensive about the viability of the project in Nagaland.

The delay in commissioning of NPPC was also stated to be a failure on the part of M/s Walchandnagar Industries Limited in erecting power boilers for the project according to the contract. The boiler tubes supplied by them started cracking during expansion of the boiler drums.

The case as regards the staff was also disappointing as the full complement of the staff necessary to run the mill could not be assembled. Raw material too was posing a problem as has been discussed in the relevant sections. In 1980, the decision to sever the two mills from HPC was mooted by the union cabinet in view of the persistent problems confronting the one at Mandya in Karnataka, and working the other at Nagaland. The centre felt that the problems could be overcome by just washing its hands off the projects and by transferring them to the state government. However, nothing seems to have materialized in this direction and NPPC is still a part of HPC.

NPPC claims that the first paper was produced in the mill on 25 September 1981, and the plant was declared as having gone into commercial production on 1 July 1982 (NPPC, write-up on NPPC, 1984).

Raw Materials and other Inputs

The mill requires 1,00,000 tonnes of bamboo and reed for the full production capacity of 33,000 mt of paper per year. The Government of Nagaland acquired 47.41 sq. miles of forest area against a proposed total acquisition of 75 sq. miles to fulfil the raw material requirement. The acquired land meets 60 per cent of the total requirement of bamboo/reed by way of extraction of 2 tonnes/acre per year. Balance requirements are proposed to be met from private cultivation by villagers in the nearby areas.

The annual requirement of other major inputs is as follows:

	Quantity required (mt)
1. Coal	1,20,000
2. Lime	16,500
3. Caustic Soda/Salt Cake	3,500
4. Chlorine	2,500
5. Rosin	330
6. Talc Powder	4,000

Due to NPPC's locational disadvantage it depends on other states for majority supply of its inputs (NPPC , write-up on NPPC, 1984).

In the Red Permanently—NPPC's Operations

NPPC was commissioned in July 1982, but for the next three years it managed to push production up to only 2,800 tonnes. In 1984-5 it sold paper worth about Rs. 4 crore which barely managed to cover expenses on raw material, chemicals, power and fuel. To earn this amount of Rs. 4 crore, NPPC spent Rs. 25 crore on fixed and variable expenses (*Update*, 1987).

NPPC had run into a state that makes a mockery of any commercial operation: the more it produces the more it loses. That happens when sale prices fail to cover even variable cost—a hopeless situation. The only reason for this mess was that the project was techno-economically doomed even on the drawing board. Two factors caused this: the realization that the Nagaland government could not run a project of this kind, and the need to put something in the hands of HPC, which had been set up with a host of lofty objectives.

Further, to cover the loss-making situation of NPPC, in a falsely optimistic and misleading report to the government on NPPC's poor performance, the HPC management laid the blame on three coal-fired boilers bought from Walchandnagar Industries. The report was signed jointly by the then Managing Director of NPPC and A. Panda, Director (Technical) who was with HPC almost since the beginning. They came up with the novel idea of replacing the coal-fired boilers with gas-fired ones. No attempt, however, was made to fix the responsibility for the failure of the boilers. HPC insiders spoke of kickbacks on these and other expensive deals to capital goods suppliers. So it was not surprising that many read similar motives in Panda's idea of installing gas-fired boilers (*Update*, 1987).

HPC's own staff was sceptical about the viability of this project. In an *Economic Times* article dated as far back as 1977, when NPPC was at a construction stage, its General Manager (Marketing) A. Chakravarti, is stated to have said that the cost of producing a tonne of paper annually worked out to Rs.13,000 against Rs. 3,000 in the early 1960s. For the Nagaland plant, this cost would be Rs. 18,000 a tonne annually. Instead of the three to four years break-even period for normal plants the break-even for the NPPC would not have been less than six or seven years. The very cost of the project is considered to be about 30 per cent higher for a plant in a place like this (*Economic Times*, 9 May 1977).

Between August 1988 and February 1989, the mill's production was completely suspended due to damage caused by a severe earthquake. Also, unstable and unreliable power supply adversely affected performance. In fact, the HPC mill in this region is now facing a complex working environment. Annual floods resulting in total disruption of movements complicate the matter even more.

NPPC could produce only 2,180 tonnes in 1988-9 against a 33,000 tonnes capacity. The mill suffered a Rs.7.50 crore operating loss in 1889-90 which was higher than the previous year's loss of

Rs. 5.11 crore. The net loss of the mill stood at Rs. 23.94 crore. After a marginal increase from 14.90 per cent in 1986-7, the capacity utilization stood at 16.04 per cent in 1987-8 but this nose-dived to 6.53 per cent in 1988-9 (*Financial Express*, 9 November 1989).

In 1989-90, NPPC was causing a great deal of worry to HPC management because of its unprofitable operations, as during the year HPC's accumulated loss was Rs. 500 crore and since inception NPPC had accounted for Rs. 144 crore. This totally wiped out the subsidiary's Rs. 48 crore equity base. This was the year when HPC achieved a record profit of Rs.72 crore. Barring NPPC all the other HPC units had made profits during the year. NPPC remained virtually non-functional and produced only 981 tonnes of paper incurring a loss of Rs. 6.89 crore in the process (*Business Standard*, 5 December 1989, and *Economic Times*, 3 November 1990).

The physical performance of all units of HPC given below indicates that there was a good pick up in production of all units except NPPC:

	1988-9 (production in tonnes)
HNL	78,660
Mandya	9,918
NPM	58,117
Cachar	37,435
NPPC	2,180

NPPC's capacity utilization went down in 1990-1, from an already abysmally low 17.2 per cent to 2.97 per cent. Besides political problems, boundary disputes, and hostilities, and erratic power supply were said to be coming in the way of NPPC's recovery. NPPC's performance was again contrary to the performance of HPC, as a whole, as for the first time HPC was doing away with budgetary support from the Union Government as non-plan cash support and repaid Rs.7.6 crore as the principal instalment (*Business and Political Observer*, 8 December 1991).

NPPC—Does Financial Reconstruction Hold the Key?

NPPC has been closed since August 1992. According to B.T. Sridharan (HPC's CMD), 'HPC was losing more by running the NPPC mill', which is plagued by power problems that have crippled its operations completely; the problem of power is due to uneasy inter-state relations between Assam and Nagaland. Now, the mill has been referred to the Board for Industrial and Financial Reconstruction, as its full worth has been eroded.

The present chief executive thinks that the question of reviving NPPC is somewhat brighter now, as a revival scheme has already been approved and is due to be submitted before the IDBI, the operating agency in this case. The Rs. 16.55 crore scheme envisages a write-off of the Rs. 60 crore loss of the mill and the government has agreed to this. However, so far, the scheme which will be vetted by IDBI, has only one major hurdle to overcome and that is to get the green signal from the Union Finance Ministry over a proposal for a five-year excise holiday. The state government has already agreed to a sales tax holiday for an equivalent number of years. Although the Finance Minister seems amenable to the idea of restarting the NPPC if it can meet its variable cost or at least a part of its fixed cost, the bureaucracy appears to be shy of approving the proposal, lest it create precedents. Till the NPPC is restarted, however, an idle workforce has to be maintained at a cost of Rs. 5 crore per annum as long as the plant is kept closed.

However, in anticipation of a favourable decision, HPC has gone ahead and signed an agreement with the Gas Authority of India Limited (GAIL) for its supply of gas from the nearby Gelli Oilfields on the border of Assam and Nagaland. This would help to solve the power problem that plagues NPPC. Commenting on this aspect, HPC's CMD, B.T. Sridharan says that the meagre requirement of the NPPC of 6 mw per day, is also not being fulfilled, as the issue of power supply is becoming entangled in the inter-state wrangles of Nagaland and Assam. Even when the Assam State Electricity Board does supply the power, it is not transmitted properly to NPPC and the Nagaland Government is not doing anything about this.

The situation is proposed to be altered through a rehabilitation scheme which envisages a Rs.7.5 crore investment for the conversion of two of the coal-fired boilers at NPPC to gas-fired ones, which would be fed from the Gelli Oilfields of ONGC. The scheme also envisages repair and overhauling of the ONGC sets, replacement of worn out machinery, and also a Rs. 1 crore investment on civil structures that would protect the plant against earthquakes as it lies in an earthquake-prone zone.

However, the scheme also envisages selling off one of the paper machines at NPPC, running the mill with half its present capacity and cutting down on the workforce by 350.

HPC sources feel sure of turning around this perennially loss-making unit to a viable one in around seven years by which time the negative worth would turn positive. Till 1992-3, the plant which was set up in 1971 and went into production had incurred a loss of Rs. 230 crore. NPPC now awaits a fresh lease of life (*PTI Corporate Trends*, 10 June 1993).

Summary and Conclusions

HPC was an afterthought. It was set up by the Government of India to produce ordinary paper as well as newsprint, since the much-pampered paper industry in the private sector had failed to fulfil its obligations and was not interested in producing a commodity like newsprint with high investment and low returns. The tragedy of HPC was the 'choice' of location. The choice in every one of the three units was to assuage the local political bigwigs rather than a sound industrial decision. The region was rejected as unworkable by all the shrewd industries. The result is a tragedy for HPC. The lush bamboo forests in both Assam and Nagaland are not the primary forests but secondary growth due to anthropogenic interference; the timber industry plays an important role. The local people have adapted themselves and manage the bamboo forest quite well, unlike the mills who are compelled to hand over the catchment area in Assam to its contractors. The bamboo protests, and the most dramatic form of protest is seen in the stunting of very tall *Bambusa hamiltoni* into short sprawling clumps. They can also vanish completely as has happened in other parts of India. It is important to remember that there is no inherent contradiction between maintenance of forests and economic development, that is, if viewed in the larger context of the National Society.

Recommendations

The Government of India must give every support to the paper mills to grow, while modern pulping technology should be introduced to save the bamboo forests. Regional development must be speeded up without destroying the forests—this will bring political stability to the region. Bamboo forests must be managed by the people and the industry should not be permitted to fell bamboo. This is an alternate in development.

A CBI enquiry must be carried out to locate and punish (a) Timber industries which have illegally felled the timber trees, (b) Those who are responsible for setting up the Nagaland Paper Industry at Tuli.

NOTES

1. The area was prospected by Balmer Lawrie and Andrew Yule and Co. in 1957 to set up a pulp mill and NCEAR was commissioned to submit a feasibility report (Assam Pulp Mill *Annual Report*). But it was given up.
2. Orissa, Madhya Pradesh and Andhra Pradesh have Birla-owned paper mills, and its rayon units are in Kerala and Karnataka.

REFERENCES

Ajoy, K. Das 1990, 'Raw Material Crunch Grips HPC', *Economic Times*, 4 May.

Alam, Khorshed 1983, 'The Emergence of Industrial Middle Class in Assam', in B.Datta Ray (ed.), *The Emergence and Role of Middle Class in North-East India*.

Allen, B.C. 1905, *Assam District Gazetteers, Cachar*, vol. I, Calcutta.

——— 1905, *Assam District Gazetteers, Naga Hills and Manipur*, vol. IX, Calcutta.

——— 1908, *The Imperial Gazetteer*, *Nowgong*, vol. XIX, 1908, Oxford University Press.

Anon, 1977, 'Cachar Paper Project may be Put Off Temporarily', *Business Standard*, 18 July.

——— 1977, 'Nagaland Paper Units Progress Tardy', *Economic Times*, 9 May.

——— 1979, 'Funds Paucity Stalls Two Paper Projects', *Financial Express*, 29 November.

——— 1979, 'HPC to Raise Share Capital', *Business Standard*, 30 October.

——— 1980, 'HPC Mills to be Handed Over to State Governments?' *Financial Express*, 10 November.

——— 1980, 'Nagaland Paper Project, Trial Runs Fixed for March', *Economic Times*, 2 December.

——— 1981, 'Engineering and Consultancy Wing Mooted for HPC', *The Times of India*, 1 September.

——— 1981, 'HPC Cuts Imported Paper Price', *Business Standard*, 18 August.

——— 1982, 'HPC's Turnover Declines', *Financial Express*, 30 June.

——— 1983, 'HPC on Recovery Path', *Financial Express*, 5 December.

——— 1985, 'Report on Laggards', *Economic Times*, 28 April.

——— 1986, 'No takers for HPC Paper', *Business Standard*, 21 January.

——— 1986, 'HPC Nowgong Paper Rs. 22.47 Crore in the Red', *Business Standard*, 12 September.

——— 1987, 'HPC Expects to Earn Profits', *Financial Express*, 27 April.
——— 1987, 'Cachar Paper Project May be Costlier', *Financial Express*, 4 July.
——— 1989, 'HPC Seeks to Snap Links with Nagaland Paper', *Economic Times*, 8 May.
——— 1989, 'HPC Won't Seek Government Funds as Performance Improves', *Business Standard*, 5 December.
——— 1989, 'HPC Optimistic about Future', *Economic Times*, 3 May.
——— 1990, 'HPC Posts Rs. 64 Crores Operating Profits', *Financial Express*, 3 May.
——— 1990, 'HP Mills Makes Profit', *The Times of India*, 7 May.
——— 1990, 'Rs. 72 Crore Profit for HPC', *Economic Times*, 13 December.
——— 1991, 'Hindustan Paper May Raise Prices', *Business Standard*, 16 September.
——— 1991, 'HPC Proves its Critics Wrong', *Business and Political Observer*, 8 February.
——— 1991, 'Panel to Evaluate Bids Soon', *Financial Express*, 1 January.
——— 1992, 'HPC Units Commissioned', *Financial Express*, 22 July.
——— 1992, 'HPC, CMD Opposes Privatization Move', *Business and Political Observer*, 1 December.
——— 1992, 'HPC Fails to Clear Huge Stock Pile Up', *The Times of India*, 30 March.
——— 1992, 'Three Point Demand to Improve HPC's Finances', *The Times of India*, 1 June.
——— 1992, 'Two HPC Units Improve Show', *Business Standard*, 22 December.
——— 1993, 'Government Dithering Delays Revival of HPC', *Business & Political Observer*, 17 June.
Bareh, H. 1970, *Gazetteer of India, Nagaland*, Kohima.
Basu, Debashis 1987, 'Another White Elephant', *Update*, 20 February.
Borr, A. 1896, *Flora of Assam: Grasses*, Calcutta.
Chaube, Shibani Kumar and Amledu Guha 1982, 'North-east India', *Social Scientist.*
Das, Amiya Kumar 1982, *Assams Agony, The Seven Sisters of the North-east,* Lancers Publisher, New Delhi.
Das, H.P. 1970, *Geography of Assam*, National Book Trust, New Delhi.
Dutta, Indrayani 1993, 'HPC: Restructuring Holds the Key', *PTI Corporate Trends,* 10 June.
Elwin, Verrier, T. Tapan Bose (eds.), 1961, *Nagaland, Shillong, Scenario of seven per cent.*
FRI 1961, *Hundred Years of Indian Forestry 1861-1961*, vol. II, Dehradun.
——— 1961, *Thirty Seven Varieties of Bamboo Growing in India*, Dehradun.
Guha, Amlendu 1980, 'Little Nationalization Turned Chauvinist', *Economic and Political Weekly,* Special Number, October.

——— 1981, 'A Historical Perspective for North-east India', *Occasional Paper No. 43*, Centre for Studies in Social Science, Calcutta.

Gupta, Shekar 1984, *A Valley Divided,* Vikas Publication, Delhi.

Hindustan Paper Corporation, *Annual Reports 1985-6, 1987-8 and 1990-1.*

Hussain, Moniru 1993, *The Assam Movement: Class Ideology and Identity,* Manak Publication, New Delhi.

Kothari, P.S. 1972, 'The Nagaland Paper Project', *Indian Forester*, vol. C, no. 8, Dehradun.

McClure, F.A. 1966, *The Bamboos, A Fresh Perspective,* Harvard University Press.

Nath, G.M. 1959, 'Flowering of the Pecha Bamboo in Cachar District, *Indian Forester*, vol. LXXXV, Dehradun.

Pingle, U., V.N. Raja Reddy, Von Furer Haimendorf C., 1982, 'Should Shifting Cultivation be Banned', *Science Today,* no. 4, reprinted in (T.K. Bose [ed.]), *Scenario of seven per cent.*

Rai, S.N. 1989, 'Tropical Rain Forests of India, Their Management and Regeneration', *Indian Forester,* vol. CXV, no. 2, Dehradun, p. 82.

Raj, Khowa S. 1961, 'Forest Types of Assam', *Indian Forester*, vol. LXXXVII, no. 9.

Ramakrishna, P.S., 'Nutrient Cycling in Forest Fallows in North-eastern India', in J.J. Proctor (ed.), *Mineral Nutrients in Tropical Forests and Savannah Eco-System.*

Ray, B. Datta 1983, *The Emergence and Role of Middle Class in North-east India,* Uppal Publishing House, New Delhi.

Sarkar, Bidyut 1977, 'HPC, Paper Giant in the Making', *Economic Times*, 27 May.

Stracey, P.D. 1954, 'Forestry in Assam', *Indian Forester*, vol. LXXXII, no. 12, Dehradun.

——— 1956, 'Development of Forestry in Assam', *Indian Forester*, vol. LXXXII, no. 12.

——— 1957, 'Rational Utilization of India's Cellulosic Raw Material', *Indian Forester*, vol. LXXXIII, no. 4.

——— 1963, 'The Forests of Nagaland', *Indian Forester*, vol. XCI, no. 11, Dehradun.

UN FAO 1954, *World Pulp and Paper Resources and Prospect*, New York.

Vinayak, M. 1986, 'Uncertain Future of HPC', *Financial Express*, 7 September.

Whitmore, T.C. 1975, *Tropical Rain Forests of the Far East*, Clarendon Press, Oxford.

CHAPTER 10

Summary and Conclusions

The Bamboo

Tree-grass unique in the plant kingdom
Symbol of uprightness, chivalry and devotion
Inspiration of poets, artists and philosophers
Writing material of the ancients,
The stylus of the contemporaries
Food, Shelter and Clothing of the people
Industrial substance of thousand uses
Redeemer of waste places . . . Protector of Soil.

F.A. McLure

TATHASTU[1]

M.S.

In this chapter, the main issues are highlighted, loose ends tied up, the hypothesis elaborated and generalizations drawn.

This project explored an uncharted area, the vast forests of India which felt the impact of PPI. The historical method proved a useful tool, a meeting point of not only natural and social sciences but some aspects of pulp and paper technology as well. This methodology helped to open out the complex scenario layered with several issues, some that the project had set out to investigate and others that were totally unanticipated but nevertheless investigated to get a fuller picture of the complex reality.

The study began by highlighting the reasons for the richness and complexity of the forest ecosystems in India and the reasons for its vulnerability (Ch. 1). History indicated that devastation began only after the entry of the European powers, but it is the British colonists who truly ravaged them with impunity till they realized the wealth that lay in the rich timber the forests yield. However, such knowledge led to enclosure of the richest areas which were 'Reserved' for themselves, ousting forest dwellers out of their ancient homelands. The forest dwellers were an integral part of the forest ecosystem. It was the beginning of the end of the richness of the forests. Inevitably, contradictions arose and began to sharpen.

British policy also created the timber merchants and timber contractors. It increased the size of the herds, reduced the cycle of slash and burn agriculture and contributed to the fuel famines. These totalled up to considerable pressures on the Indian forests during the colonial period. And then came the final loot for the war effort when the colonizers knew they would have to quit (Ch. 4). They did quit, but they also left behind their legacy of various pressures on the forests.

Raw material for paper is only one of the pressures, and it need not have been so as bamboo was a hardy undergrowth and a truly renewable and, in fact, a self-perpetuating resource.[2]

So, the main research questions were: To what extent is the destruction of forests due to the growth of the PPI and what is the role of the state in promoting the PPI since production of paper is an index of development. How does the contradiction between development and environment manifest itself in the specific case of the PPI and forests, what policy recommendation can be derived from it, and what alternates in development can be suggested as a part of the IDPAD programme?

The study indicates the British colonizers' earliest sorties were on the forests on the Western Ghats, and so too, their desperate measures to set up plantations (Chs. 6 and 7). The tribal question, their relationship with their forests and their resistance to British rule appear mainly in Chs. 2 and 9, the debate on slash and burn agriculture in Chs. 4 and 9. The colonizer's desperate need for fuel to keep the railways moving manifested itself clearly in the Madras forests and the forest departments which were geared to this need (Ch. 7). In the turbulent colonial period, railways were primarily meant for troop movements and the movement of cotton. Wellington was in the Nilgiris, where the white soldiers recouped after the heat of numerous battles and the heat of the tropical sun. While some of the pressure that the colonial government unleashed on the Indian forests remains, our concern is with the pressure created by the pulp and paper industry. During the course of our study we have raised certain issues, for instance in the twentieth century backward economy—who should have priority for bamboo and reed, a newsprint manufacturing unit or the 'mat-makers and basket weavers' (Ch. 7). FAO (1996: 18) belatedly pays lip service to the rights of people for the forest produce.

The transformation of a section of the Indian merchant class, and its struggle to rise to the status of a powerful industrial class and as paper giants, is explained in the introductory chapter, illustrated perhaps most cogently through the rise of Birla and J.K. Singhania (Chs. 1, 2), the astuteness of this class by a somewhat late entrant, an American trained Thapar (Ch. 4) and to a lesser extent, by Bangur who displayed his old mercantile mentality (Chs. 5, 6). The manipulative power of international capital which has ganged up with the Indian paper cartel and even claims to have assumed the leadership to browbeat the government is found in Ch. 5, Section 3 ITCB. A combination of clout and cunning, but for a totally different reason, is displayed by TNPL in which the World Bank holds 12 per cent equity (Ch. 8). But, by and large, the huge forest based paper and rayon industry in India is controlled by four of the biggest business houses. The entry of Seshasayee is incidental, while the entry of the international ones is a peculiar attempt by the southern states to keep Marwari capital out.

Broadly speaking, the destruction of the bamboo forests is almost total. There are, of course, variations in the degrees of destruction. The Birlas, both the paper as well as the rayon branches of the divided family, can claim the dubious distinction of being the most rapacious. In both Assam and in Nagaland, the Birlas have timber interests. The total destruction of timber trees in the north-east has resulted in the growth of secondary bamboo forests (Ch. 9). In other words, except in Tamil Nadu, the Birlas have destroyed the Indian forests beyond redemption.

What emerges very strongly from the study is that the state is biased in favour of the private sector PPI, to the detriment of the natural forests and the forest dwellers. Birla was one of the planners for India's mode of indicative planning for the development of a new nation. Perhaps because of this important antecedent, Birla, more than any other owner of a PPI unit, was able to manipulate the state structure for his own benefit and sometimes to the disgust of other PPI units (Chs. 6 and 7). More importantly the forests were the traditional home of the adivasis, the tribes.

It is necessary to go back to the question of the forest dwellers and their right to their forests which the British chose to enclose as 'Reserve Forests' for their rich timber, ousting the tribals. The forest dwellers

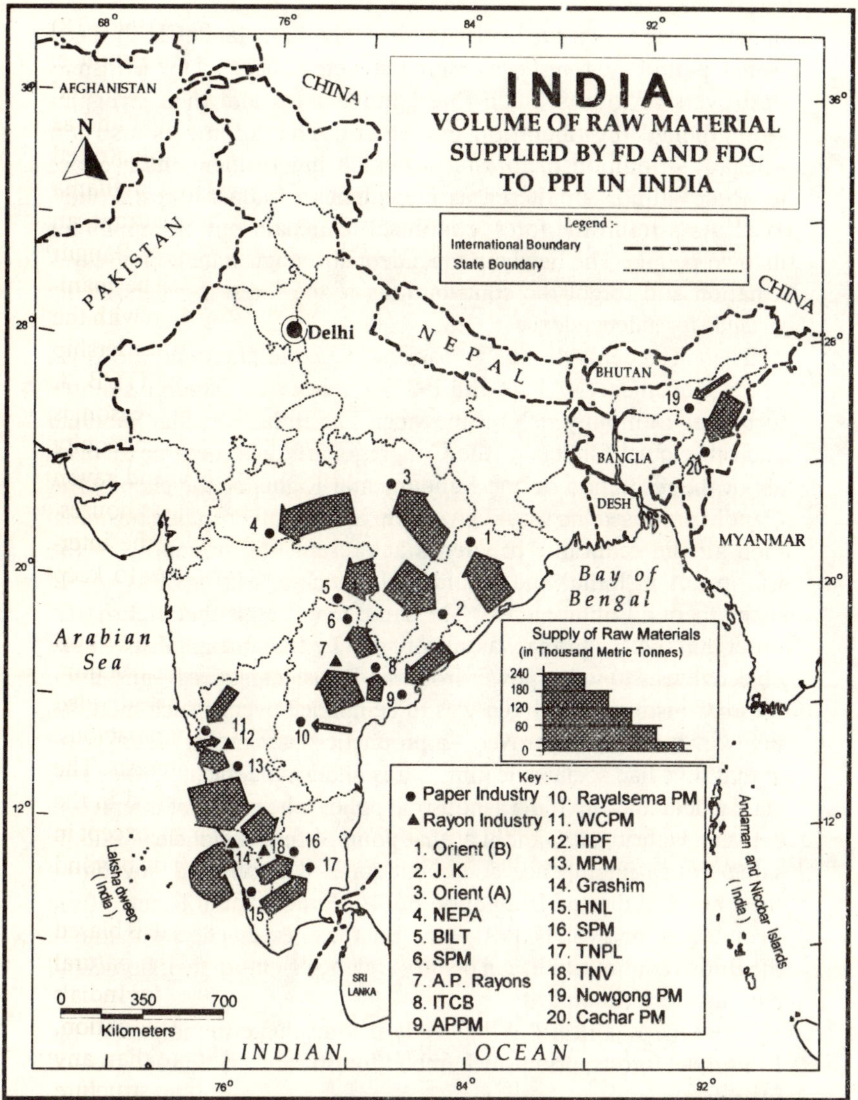

Index to Places Numbered in the Key :

Map 10.1

1. Brajrajnagar	6. Sirpur	11. Dandeli	16. Erode
2. Rayagada	7. Warangal Dist.	12. Harihar	17. Pugalur
3. Amlai	8. Bhadrachalam	13. Bhadravati	18. Mettur
4. Nepa Nagar	9. Rajahmundry	14. Mavoor	19. Nowgong
5. Ballarpur	10. Rayalasema	15. Vellor	20. Cachar

in most areas fought the British but lost their homeland (Ch. 2). They were a part of the forest ecosystem but were dubbed by the British as destroyers of forests. Such forest movements and their struggles strengthened the liberation movement. The colonizers also left a legacy—tradition of forestery—which independent India chose to accept almost unaltered as is evident in India's Forest Policy, 1952. As a result the forest continued to remain out of bounds for its own people. The newly-independent sovereign state forgot it was a nation and forgot the contributions of the forest dwellers in the struggle for independence.

The antecedents for this go back to 1931 and are found in British Secret Document (NAI, 7 April 1931). It reports on Gandhiji's simple technique to mollify rich businessmen like Birla, Ambalal Sarabhai and other quiet funders of the Congress Party who became anxious about the resolution on the Fundamental Rights of the people. But Gandhi promised 'he would give them and the other Indian businessmen all the economic privileges at present the monopoly of the Europeans in India' and would build 'such a tariff wall as would make them all millionaires'. The same report adds that Nehru was upset that the Mahatma was surrounded by Indian capitalists whose objective was to wrest power from the British capitalists—and if the future Constitution of India was to be handed over to such men, his life work would be destroyed—a prophetic statement by Nehru, who at that time had socialist leanings. It is understandable, that the same class would draw up India's industrial policy. The nature of interaction between Nehru and Gandhi at that point of time is not clear, but the same British intelligence document states 'Gandhi and Jawaharlal sat down and drafted Resolution no. 16, Fundamental Rights of the People' (Chopra, vol. II, 1991: 145). But the forest dwellers who fought the British continue to live in penury and their rights to their homeland have not been restored.

The central concern of this study is to evaluate the impact of the PPI on the forests, primarily bamboo forests, and on 'forests' created for PPI.

A hypothesis was stated in the first chapter and a corollary added to it. Both emerged out of the observations made in the nine states, and the findings presented in the substantive chapters. These observations are counter checked with literature on the subject and compared with the earlier status of forests as stated in the nineteenth century *District Gazetteers* as well as with the new series. The latter provide information till the 1960s of this century. Here the hypothesis will be elaborated and certain generalizations made. We suggest that both the Bamboo Centre at the Kerala Forest Research Institute as well as the Indian Council of Forest Research and Education at its Coimbatore and Jabalpur units test the hypothesis in order to bring bamboo back as a worthwhile raw material primarily for production of printing and writing paper as well as newsprint.

The hypothesis refers to the death of bamboo forests in the catchment areas of every one of the PPI units which is an amazing feature, considering that bamboo belongs to the family of grasses and has a dual system of reproduction. To briefly recapitulate: Bamboo's vegetative growth and spread of the rhizome goes on unfailingly from 20 to 80 good monsoons depending on the species. After its regular physiological lifespan it flowers gregariously, and the seeds, enriched through cross pollination, scatter and the old clumps die to make room for the new. The winds spread the seed and the seedlings sprout with the first showers. In the initial stage the seedlings are indistinguishable from succulent grass and cattle browse on them. It takes eight to ten years for bamboo to establish itself as a healthy clump and the cycle goes on. But this cycle was repeatedly broken by PPI as the project data indicates.

Bamboo is so aggressive a species that it earned the epithet 'the despicable weed' from the colonial foresters when it often invaded their precious teak plantations and smothered the teak saplings. And for the same reason the useful bamboo was always planted at the far end of the compound of wealthy agricultural households. Bamboo is so resilient that it withstands immense pressures, the seedlings are browsable, its culms are elephant feed, the mature bamboo provides livelihood to bamboo artisans, small 'timber' to villagers, and has through its sheer bulk once yielded enormous revenue to the colonial Forest Department. This even surpassed the revenue earned by expensive timber like teak; and from 1925 it provided raw material for PPI without bursting into flower till the late 1950s.

In the post-Independence period, occasionally in the 1950s, the output of bamboo began to fall and it rapidly went down to 1 tonne per acre by the mid-1960s when the potential with good management in natural forests could have been 4 to 6 tonnes (FAO 1966: 144, quoted in *Economic and Scientific Research Foundation*, [ESRF] 1969: 28). This fall was due to the arbitrary decision of the central government to pitch the royalty rate of bamboo so low that it yielded little revenue for the state Forest Departments to invest and manage their natural forests to yield their potential. The foresters were angry at the new industry-state linkages which not only bled the forests, but again did so with the introduction of eucalyptus in the forests as the potential raw material. The foresters were not consulted. The State Forest Departments had neither the resources nor the mandate to nurture the forests and bamboo that once grew so luxuriantly as an undergrowth. But the industry on the other hand was so pleased, that it turned reckless and sapped the sustaining power of the bamboo, so that the new units which entered the PPI scenario were mainly based in areas with tall brambly *Bambusa arundinacea.* The culms of these tend to get tangled at the upper end and it is laborious and time-consuming to follow the felling rules laid down in the working plan. For the industry 'time is money' and, possibly with eucalyptus as an alternate raw material in the offing, to felling rules were ignored. *D. strictus* also grew in the same forests and it is easy to apply the felling rules to this species. But they were not, and PPI made a clean sweep of both the bamboo species in the immediate periphery of the mills and then penetrated deeper and deeper into virgin forests to ravage the raw material (Chs. 6, 7). Bamboos reacted in these regions and began to flower gregariously within two years of setting up of a PPI unit. The time-honoured physiological cycle turned away; from the 40 years of *D. strictus* and 80 years of *B. arundinacea* this cycle dropped for both to around 10 years, of which 8 years is the time period assigned for growth after immediate resurrection through seeding. Our hypothesis is that a combination of overfelling and wrong felling practices gave rise to this defensive reaction, gregarious flowering and death. This hypothesis is based on coalescence of anthropological information with ecological and botanical assumptions. Anthropological studies particularly of the central and north-east tribal belt report the great anxiety of tribals, the forest dwellers, when bamboo flowers. They bewail it, as it portends severe famine. It is important to remember that bamboo grows rapidly to its full height anywhere from 20 to 80 ft in a brief three-month period during the rains, at about 1 ft a night. The monsoon in India hardly ever fails, and if it does, it is for two or rather three consecutive years which of course, results in famine, as agriculture by and large is rain fed; in the latter year famines are more distressing than in the earlier, as there would still be roots and fruits and kernels in the jungles for survival in the initial year. Cereals are rainfed crops and fail to yield if the monsoon fails. Cereals are cultivated grasses while bamboo is grass rigid and woody with lignin but essentially a cellulose, a carbohydrate $C_{12}H_{22}O_{11}$ with a large water component. Good rainfall, or running water (i.e. stream) are essential for survival and growth of bamboo. Water is both an agent of nutrient distribution in nature and a significant medium of biological activity in plants. The repeated failure of the monsoon rings the death knell for bamboo. A defensive reaction is seen in that, as a means of the survival of that species, bamboo flowers gregariously before it succumbs to abiotic pressure and dies. In distress, time perspective goes askew and that is the plausible reason why bamboo is reported to seed gregariously in famine years (Kadambi 1949: 295).That an enormous quantity of its seed is collected cooked into a delicious meal in famine years is another matter altogether. The rhizomes of such gregariously flowering bamboo is reported to be thin and emaciated. In other words stress beyond its tolerance point whether abiotic or biotic appears to be the causative factor for gregarious flowering. Dr. Kadambi, a silviculturist agrees with our argument on water stress (Kadambi 1949: 249). One of the major stress factors we hypothesized is a combination of wrong felling and overfelling by the mills in addition to the unnaturally short three or four years recommended felling cycle.

Again the felled culms are taken away, which also implies that the natural nutrient cycle is not only cut but that the nutrients are drained out. Cleaning of the selectively-felled clump is always a working plan prescription. This again cuts into the food chain and further impoverishes the soil and the potential nutrient intake.

Even if it is assumed, for sheer convenience alone, that felling practices laid down by the British foresters were correct then in every case it was found that the rules are flouted, particularly damaging being the tendency to fell the culms in the outer most circle which must be left intact to provide nutrient for the rhizome beneath to spread outwards. Otherwise, this turns inwards and gets tangled as there is little space to grow. Secondly, failure to cut the culm clean, leaves it jagged and splintered. So, the hypothesis gains further strength from the observations made by so careful a student of Indian forests as R.S. Troupe, Professor of Forestry at Oxford University, and a respected member of the Indian Forest Service.

Troupe (1911: 981) frequently observed a profuse mass of flowers appearing on the stump of the cut culms and on small shoots issuing from the base of the clump in areas cleared for temporary cultivation, particularly in the case of *Bambusa tulda, D. strictus* and *Cephalostachym pergracile.* Although the physiological life-cycle of a bamboo is inherent in the plant, Gamble (pp. 31-3) mentions cases of clumps of *B. tulda* and *B. nutans* producing flowers before their time when injured by cutting or partial uprooting. Troupe also thinks that 'Mutilation may, as in the case of fruit trees, stimulate flowering' (p. 982). The prevalence of dense bamboo forests in both Assam and Nagaland confirms the team's hypothesis. In Assam most amboo is grown in homesteads and harvested by the owners, forests are owned and controlled by the community. Neither the Forest Department nor the mills have a right to fell bamboo in Nagaland (Ch. 9). A third observation which further strengthens our hypothesis is the resurrection of bamboo in Karnataka after a ban by the PPI was imposed on felling it. the bamboo artisans are permitted to cut a bamboo or two for their need. In anticipation of exhaustion of the bamboo forests, various species of eucalyptus were introduced as plantation crops in the most potentially fertile forests all over India. Officially, it began with 1 per cent in the third plan and raised to 2 per cent in the fifth plan and the natural forests were clear-felled for the purpose. The plantation programme began in 1955/6 in Madhya Pradesh and in Kerala in 1958/9 without trial, merely on the advice of FAO, and this activity sent wrong signals to the PPI (Chs. 3, 7). This is the corollary to the main hypothesis. Much later the other exotics were introduced as the flood gates were opened. FAO and other international organizations like WB, SIDA, ODA, besides FAO, were ready to participate in forestry programmes with advice and loans.

FAO has belatedly woken up to the virtues of bamboo and claims it supports micro-propagation of bamboo developed by scientists in Asian countries (FAO, 1 Oct. 1993). *Forestry Topics Reports*, no. 4, p. 24, but the *Directory of Bamboo Researchers of South and South-East Asia* 1994, published by Bamboo Information Centre, Peechi, India, does not confirm FAO's claim with one exception, viz., ERDB/DENR/UNDP/FAO had organized a symposium on Bamboo Research and Development in Laguna in April 1992.

The first question that remains unanswered is why should exotics have been introduced without any trial and on such a massive scale? Thirawat, a FAO team member to Australia in 1952 and a thai Forester had sounded alarm bell for India in 1955 and again in 1956 (Ch. 1). India paid no heed to it although Thirawat's arguments were based on his observations in Australia, the struggle in Thailand (which shares the Indo–Malaysian forest ecosystem) to establish eucalyptus plantations, as well as on all available literature relating to India, frantic attempts made by the British in India to cultivate eucalyptus since 1840, and its failure to acclimatize. The Nilgiris, he states, with its unusual topography and climate was an exception where *E. globulus* thrives (Thirawat 1955: 661). For some strange reason the Indian government preferred to accept another FAO expert J. Von Monroy's reckless advice. He was an Austrian forester said to have been much impressed by the Australian species but with little knowledge of either Australian or Indian conditions, reported a couple of senior foresters of the 1958-9 batch from FRI who had met him. The second question is why India failed to keep track of FAO reports on its experiments published from 1950-73 (FAO 1980: 1). The final two-volume consolidated report gives information both on conditions under which specific species grow as well as on their pulpable qualities (FAO, 1980). And FAO under the terms of its charter bore 'the responsibility of furthering action to ensure the output of the forests in quantity and quality rise' (FAO, 1954: v) in which it failed.

The poor response of the main species *E. tereticornis* perhaps induced Government of India to ask FAO to review the species in 1970. The findings of this expensive negative report, both on its pulp and growth quality (Ch. 8), were not communicated to the State Forest Departments and the Forest Development Corporations. The third question is who in New Delhi blocked the passage of this vital information? Will an enquiry be conducted to pin down the person responsible, the person who continued to drain the exchequer, pollute the rivers with the non-volatilizable black liquor of *E. tereticornis* which continues to cause suffering to the people. It also causes loss of fertility to the farmlands where the species are cultivated as farm forestry programme and promoted by FAO as a part of sustainable development. FAO (1993) and agro forestry (FAO, 8-21 September 1995: 6) should the blame be laid only on two indifferent impersonal bureaucracies, those at Rome and at New Delhi? Theoretically viewed, do we then discard Max Weber and his theory that bureaucracy is co-equal with efficiency and an epitome of the highest western rationality? There is no information whether these hardwoods that grow in Australia and used as cellulosic raw material for paper also yield non-volatilizable black liquor, and if so, how does the industry in that country cope with it?

The fourth question is why was eucalyptus recommended by FAO and supported by the WB in the Social Forestry Programme? The data points to the National Commission of Agriculture (1976) both acting as a front and merely echoing it. Surely, it was the simplest thing for an array of foresters from several nations to realize that the evolution of flora and fauna is totally different in Australia which was cut off at a very early stage of evolution of the species from the rest of the world. The symbiotic relations that are established between eucalyptus and other native flora and fauna in Australia does not hold good for India. In Australia, the Koala bear and the birds roost on the eucalyptus, while bees collect their honey from eucalyptus flowers and many of the species have a rich undergrowth. But this does not warrant perpetuation of such a species in India. Here, no birds, no bees, even the perpetually hungry goat or the lowly weed cares to establish any symbiosis with eucalyptus. It is not that the tall, stately and truly handsome species should be despised. It is clear from an evolutionary point of view, that the antibiosis of eucalyptus is merely the result of metabolic left-over or allelochemicals, which are of no use to the species; yet in the struggle for the survival of the fittest they must have served that purpose, which perhaps explains why 96 per cent of the trees in Australia belong to the vast family of eucalyptus.

The metabolic wastes, the allelochemicals which modify the growth and behaviour of other plant organisms and microbes, benefit the parent plant by inhibiting the growth and germination of many other species of plants. J. Kimmins (1987: 367) reports that in North America no other species survives along with the stands of eucalyptus, although the species in its native Australia, has a well-developed under storey, perhaps because the evolutionary processes in the region are different. Evolution has failed to find tune India's native plants to the bio-chemical wastes of the exotics and are therefore endangered by them. Eucalyptus thrives in Brazil and the southern parts of Africa but there is no information about its effects on the local flora and fauna.

Buchanan (1938: 19-96) quoted in Kimmins (1987: 368) thinks that the soil too, has a role to play. Thirawat highlights the differential of soil on varieties of eucalyptus in native Australia (1955: 661-5). The innocent buttercup that grows happily along with other species in Europe, and which caught the fancy of the romantic poets of nineteenth-century England, turned allelopathic and excluded all other species in Utah (USA), especially in sandy soils, but less so in fine textured soils, presumably because allelochemicals were absorbed and rendered ineffective in the latter soils. It is now generally accepted that allelochemicals are more prevalent than was generally thought; what needs to be avoided is only a mismatch between species and species, as well as species and sites, whether they be degraded forest lands (Ch. 6), or farmlands (Ch. 5). The non-degradability of leaves of certain other exotic species, again Australian in origin, (Ch. 6), has perhaps something to do with the Indian soils too. Most of the questionable exotic pulpable species are Australian in origin, a country also known for its bush fires. The role of the bush fires in degrading the non-degradable 'leaves' as well as in volatilizing the essential oils from other types of leaves and pulling them back into the food chain is still not clear, perhaps because it has not been investigated.

FAO seems to have erred in many ways through its incompetence as well as lack of concern. Apart from recommending eucalyptus as a pulpable species, it also recommended wrong sites for setting up PPI units (Chs. 6, 7, 8, 9). It recommended pulpwood plantations holding Sweden as a model (FAO, 1968), but totally ignoring, Sweden's natural and social history. It knows that there is a tremendous amount of land hunger and poverty in India, a similar situation in mid-nineteenth century in Sweden when a large section of its population emigrated to US and then the pressure on its forests also eased. Sweden's forests are private or cooperatively held and the state cannot 'gift' away forest produce to the industry as in India. On the contrary, saw mills were banned from acquiring forest lands as early as 1904. World War II did not damage the Swedish forest, but the rising demand was satisfied by the import of round wood and processed wood. Subsidies combined with extension law and supervision aided the improving of Sweden's forest resources. Resources generated by industrial forestry are used for conservation and regeneration, something which is unknown in India. The nearest valid comparison is Nagaland forests but the rest of India's forests do not compare even with Nagaland. Although in 1989, FAO formally admitted plantation as a monoculture being a danger, unless closely monitored, it persists in recommending farm forestry to create pulpwood for PPI. FAO is thus resorting to double standards. It has made no efforts to retract its position of cultivating eucalyptus in India and some other tropical countries where eucalyptus fails to yield. There is no information of eucalyptus effects on the soils in other countries. It is only with the establishment of FRI that a document routinely listing the negative features of eucalyptus has been published, and the social content of the social forestry programmes questioned, even though mildly. The entry of the international bodies was so insidious that the damage done was not seriously questioned right up till the1980s. The foresters in the state forest departments were by and large so hamstrung by the bureaucracy that they did not wish to discuss the issue. There were some who quietly fought against it (Ch. 4), but did not speak about it even to us.

The damage caused by the exotic species cultivated to feed the PPI has gone beyond the forests. PPI's units have extended their catchment areas into farmlands. Destruction of the soils in the farmlands, most often those of poor peasants unwillingly drawn into cultivation of pulpwood exotics, added an unanticipated dimension to this study (Ch. 5). It is surprising that many enviromnentally sensitive and ecologically concerned individuals, by and large based in Delhi and Western UP, should plead for a tie-up between the paper industry and the farmer, when the latter cultivates the same dangerous eucalyptus species (Saksena 1990). Saksena has since changed his stand. Land relations in Punjab, Haryana and Western UP namely the 'green revolution' belt, are atypical of India, especially since post-colonial India did not flilly carry out land reforms. Reform did begin, but only due to pressure by the powerful feudal lobby, ending in mere lopsided tenancy legislation. Poor peasants caught in it became victims (Ch. 5). The suggestion therefore made by those concerned are short-term measures to save the natural forests, but at the expense of both the farmer and the farmlands. Only land reform can save the situation. The first alternative of course is revival of the bamboo forests and encouragement of homestead bamboos where soil and climatic conditions are suitable. This tradition is still prevalent in a minor way in Kerala, Orissa and Bengal and in a big way in the north-east. The insatiable paper industry is fed by homestead and community bamboo in the north-east, yet bamboo thrives, which only confirms our hypothesis that careless over-felling creates stress and has almost wiped out bamboos from the face of the Indian forests. The FAO/Indian team (1980) rate *Ochelandra* species of bamboo superior to temperate soft woods. But these grow only in wet evergreen forests at higher altitude.

There is no shortage of safe raw material which will not harm the ecosystem. There are many types of grasses, and straw—some of these are cattle feed but others like jute, hemp, *kenaf* are native and cultivable.

Faced with a raw material famine, the paper industry in the 1970s concentrated its R and D on finding alternate sources of raw material, which it discovered in the *Hibiscus* family.

Hibiscus mutabil, popularly referred to as *mesta* in India and *kenaf* in US, is a native of Andhra Pradesh. It is a cousin of hemp, jute and cotton that grows in many regions in India. Hemp is rated the highest for its cellulosic content, but is said to be too expensive and unremunerative for PPI (Ch. 2). The strength value of *mesta* pulp is comparable or even higher than that of bamboo pulp. The stated reason for not using it is its lower bulk density. But the industry's own R and D has shown that this can be overcome by making proper bales in the fields, and by installing a continuous digester equipped with a pressure impregnation system. *Mesta* pulp also bleaches more easily by CEH sequence at a lower bleach requirement as compared to bamboo. Since its silica content in *mesta* black liquor is very low, it makes it favourable for black liquor evaporation (Saraf et al. 1980: 31). In other words, the use of *mesta* implies substantial investment in new digester, but, in return would yield a good raw material with an appreciably lesser degree of pollutant.

The question of disposal of lignin remains still unresolved. The present Managing Director of TNPL declared *mesta* to be the most promising raw material for both large and small paper mills, as well as for NP (Tushu and Murthy 1991: 100, 105). Hemp as well as jute and cotton stocks have also been investigated and approved as raw material. Why then is there no tie-up between the farmers and the industry to grow this eco-friendly cellulosic raw material? It is important to remember that the mega industry thrives on large margins of profits. Its bargaining power with farmers who cultivate short-term crops is considerably reduced. Farmers can easily change their cropping pattern and unlike the state cannot give the raw material at a subsidized price. They are known to have hiked the price of both *mesta* and hemp as soon as they found out that their customer was an affluent paper mill. The votaries of a free market should not stop. The farmers need the government's support. In 1985 the West Coast Paper Mill roped in the local peasantry, provided them with inputs and seeds, bought their entire output of *mesta* and subtracted the cost of inputs. It was an experiment that failed. The reasons for failure are not clear. But what is clear is that the industry is unwilling to share the fruit of its growth with the producer of its precious raw material. There is now a dangerous trend, varying from persuasion to profitable deals with the peasantry to use effluents to grow raw materials, which poses a threat to the fertility of the soil and pollutes both groundwater as well as underground water. NEERI needs to put in greater effort to compel the PPI to maintain safe standards. The Ministry of Environment concurs.

It is not merely the new raw material that saves the forest product but also the type of technology used. The more advanced this technology, the larger is the yield of cellulosic material and, proportionately, the lesser the pressure on the ecosystem. Compulsion to use advanced technology is also a product of history, not merely temporal but also social, brought in by pressures from the ecologically conscious group to conserve nature. The yield of cellulosic raw material can vary from 33 per cent to 90 per cent. The latest method of pulping is the oxygen delignification process. Indian private sector units, without exception, use the obsolete and the most wasteful sulphate process. Any discussion on this question with the representatives of the industry ends on the same note, which is that investment on a new pulping machine, continuous digester, etc., is too heavy for the industry to bear. The standard counter-argument for long-term gains of sustained yield of critical raw material and the question of enlightened self-interest is dismissed by the simple yet truthful answer that the paper industry is a business and the top management's function is to make it yield profits. The PPI looks upon enlightened self-interest to save forests for sustained yield as a purely academic question. Introduction of eucalyptus has made it impossible to use the highest material saving technology, viz., thermo-mechanical pulping a chemical component has to be added which brings down the yield from 95 per cent to 85-90 per cent and it is only utilized in the public and joint sector newsprint mills, thanks to the Indian FAO team which worked on it from the mid-1970s to 1980s. Desperation because of dwindling raw material and unwillingness to compel the private sector to modernize technology to produce socially useful cultural paper and newsprint is responsible for this. Disposal of lignin was a problem; it is the only known non-biodegradable organic substance and was used as a landfill. But now an oxygen delignification process is available.

There is no reason to believe that the western bourgeoisie through time has become more thoughtful and does not destroy its own resource base. A capitalist's concern is profit, here and now. A capitalist cannot wait. If the paper industry in the west has upgraded its technology, cut down consumption of its forest-based raw material or reduced the levels of pollution, it is not due to enlightened self-interest, but because the industry has been compelled by incessant pressures from various environmentally conscious groups to do so, right from the late 1960s in the case of the USA. The foresters in the state service in that country are also products of the same powerful movement and now zealously guard their forests (Grey Beal, senior official at the Oregon State Forest Department, Sinlow National Forests, Personal communication, June 1986). But their interest in environmental preservation is parochially narrow.[3]

Today there is tremendous pressure on the government by the PPI to release land for its captive plantations of eucalyptus. Information on negative qualities of eucalyptus are not made available to the industry. Earlier, the National Commission of Agriculture (1976) had debated this issue at length and formally rejected it wholly. The question has been raised again after two decades. The foresters in the Ministry are stoutly opposing it, but are unsure of the outcome. The threat is real, it is proposed by National Forest Action Programme (NFAP) (*Update* 32: 21). Yet in 1992, in Orissa, there was a combined move by the most powerful units to set up a joint sector management of pulpwood plantations and the Chief Minister had accepted the proposal (Ch. 2). The Forest Development Corporation is terrified. They fear a dangerously potent combination of an extra load of work as well as loot. The state Forest Departments which are compelled to raise pulpwood plantations at prohibitive costs are also compelled by old agreements to sell the outturn at below the cost of raising the trees (Ch. 8). Any attempt by the Forest Department to raise the royalty rate to match the cost of raising the raw material is challenged by the industry through a court of law.

Modern 'justice' which goes by the letter of agreement, however unfair it may be, is blind to the changed social reality, additionally daunting the foresters. Foresters appointed as guardians of the forests are caught in this bureaucratic web and are stifled by political dictates.

Only West Coast Paper Mills which had recklessly destroyed its leased forest areas, tried to make amends by underplanting bamboos within its forests. But the unit became possessive and enclosed the forests with triple barbed wire and a moat. The local people, denied their fundamental right to 'minor' forest produce—twigs as fuel and grass as fodder for their cattle, naturally reacted. The unit lost interest in the revival of the bamboo forests whose owner Somani, was soon tricked out of his rights to the West Coast Paper Mill by his financier. He has now asked the government for ravine land to grow his own raw material for his mill in south Gujarat (personal communication, N.K. Somani, 1 June 1994). At face value, this demand is innocent, but if the government gives in to it, the other paper units will ask for similar grants. The line between wasteland and forest land is thin and the lines can always be redrawn (Ch. 2) to suit the interest of the industry. The industry has established its reputation as a rapacious grabber and destroyer of forests. The fears of the foresters are very real. The need to save the forests, the very life support system, has to be given priority. Paper and NP is also vital for development but the industry has alternatives, and the alternatives will be produced only if the producer of the safe raw material gets a fair deal.

What emerges very strongly from the study is that the state is biased in favour of the private sector PPI to the detriment of the natural forests and lack of concern for its own public sector paper and newsprint units which are way ahead of the private corporate sector. Birla's paper units, for instance, in both Orissa and Andhra Pradesh are in a totally dilapidated condition the chief executive of its Madhya Pradesh unit repeatedly stalled the visit of the team to this unit. The people's movement in Kerala has compelled the closure of Birla's Rayon Pulp Factory (Ch. 7), and the people's movement Karnataka has resulted in a ban on felling by the industry and closure the joint sector eucalyptus plantation (Ch. 6). As one senior forester in the Ministry of Forests said 'The people's movement will save the forests in India.' So be it.

This study on the impact of the PPI in Indian Forests has revealed not only the death of bamboo forests but also unexpected issues, such as the resistance of farmers to extension of plantation 'forests' into farm lands and village common lands with ecologically damaging species. It also reveals the unexpressed resentment of foresters to the centre's high-handed ways in which the Ministry of Environment and Forests (MEF) is a mere front, and the attempt by the Government of India and FAO to keep under wraps the failure of the commonest eucalyptus species, viz., *E. tereticornis* as not only ecologically unsuitable but also as an inappropriate cellulosic raw material for PPI. One may under these circumstances ask: Can FAO be considered a knowledgeable organization for India to be dependent on for advice and collaboration?

FAO is aware of every one of the issues which have been raised. It tried to deal with them by merely making cosmetic changes in the original policy which the Government of India carried out, but to no avail of course. Under its heightened awareness of the Indian situation, as well as of those in the other countries, FAO claims to have broadened its viewpoint and it no longer considers forests merely as raw material producers for industry (as it did in the *National Commission of Agriculture*, vol. IX, 1976: 1, 4) but as complex ecosystems which provide essential goods and services for local people, also as the main prop for environmental stability. With this heightened sensibility it has offered a National Forestry Programme (NFP) for third world countries to be carried out under its guidance. In India it threatened to come in 1997 as National Forestry Action Programme (NFAP).

FAO in its revised NFP (1966: 14) clearly states 'the *goal* of NFP is to promote the conservation and sustainable use of forest resources to meet' not only 'local and national but also global needs, through fostering national and international partnerships to manage, protect and restore forest resources and land, for benefit of present and future generations. Why global needs should be met by mismanaged and depleted tropical forests has not been clarified.

In order to achieve NFP's goal its first 'basic principle' is 'sustainability of forest development' and the second principle assigns 'full leadership and responsibility to the National Sovereign Country' (1996: 22) for carrying out the onerous task. Sustainable development of course is now recognized as a deceptively simple catchword which does not function[4] in an inequitous world economy (Leslie, FAO 1994: 20, 25). International commitment is listed as the twelfth and last basic priinciple which exhorts the international institutions to 'adapt their own priorities to the (host) country's priority' which priorities have not been spelt out nor the reason why tropical forest programmes should fulfil global needs and what precisely these needs are. They were brazenly announced when FAO made its first forays nearly half a century ago.

India had very quietly entered National Forestry Action Programme in September 1993, with UNDP/FAO as the core support agency.[5] It did so without any publicity or debates. The exercise was confidential and some NGO participated in the preparation of the technical or specialized report says *NFAP Update* 32 (pp. 18, 19). FAO realizing the contradictions it had created between the State Forest Departments and the centre with NCA (1976) as the mouth-piece had resulted 'in earmarking of 70 to 80 per cent' of budgetary allocation for social forestry. The upshot was 'the establishment of high value plantations at the expense of natural forests resulting in loss of bio-diversity and non-wood forest products. As a result there was opposition to the practice from the people and the government had to revise the plantation programme strategy.' In order to ensure that NFAP be country driven, state Forest Departments were assigned the task of preparing State FAP (NSAP *Update*, 32: 19) but the latest available *Travel Report, India, 8-12 September 1995* dated 28 September 1995 (p. 3) reports that none of the states have so far submitted SFAP agroforestry on the agenda (p. 6). So too the dangerous proposition of opening out degraded forests to industry to raise captive plantations (*NFAP Update*, 32: 21). The final draft of NFAP is expected to be completed in 1996 and implemented in 1997 (*NFAP Update*, 32: 27). In other words there is no radical change in the Government of India-FAO position except that the legal responsibility will be fully India's. Does India with trained foresters, silviculturists, ecologists, anthro-pologists, highly developed *Ayurvedic* and *Unani* as well as folk and tribal systems of medicine based on jungle roots, herbs, bark, kernels and minerals,[6] capable scientists, chemical engineers, paper technologists and shrewd and sophisticated industrialists really need FAO's advice again?

The above question is seriously raised mainly because under the growing pressure of Western capitalism the Indian state has succumbed and opened out vast spaces for the international bourgeoisie especially to America for its goods, services, capital and technology. Indian industry has foolishly paved the way for it by clamouring against what it called the 'license raj'. The Indian paper giants are terrified. Their reactions are presented, perhaps, most clearly in Ch. 4. Even the Indian Tobacco Company was reacting strongly against the move by its parent company (Ch. 5). Cheap labour, the possibility of raising raw material if forests fail, lack of stringency in rules, a pliant government, a potentially vast market, are attractive avenues through which international paper giants can enter. The public sector units, which already had a raw deal, have been the first victims (Ch. 9). The ecologically friendly and raw material-saving public sector newsprint units are threatened (Ch. 7), due to the short-sighted policy of the powerful newspaper industry and newsprint dealers, who in collusion with the Ministry of Information and Broadcasting, have created havoc for the producers of newsprint.

Multilateral agencies, FAO, SIDA, ODA, USAID and the World Bank (Ch. 3) have played a negative role by introducing exotic species in Indian forests.[7] Their spread into degraded forest land (Ch. 6) and in agricultural lands is a grave threat, both to the ecosystem and to the poor peasantry (Ch. 9), especially in a country where agrarian reforms have not been carried through. Will international capital attempt to give these mischievous agencies another lease to create trouble? Will the people's power gear up to thwart such a move? Eventually, this is what should happen.

One of the objectives of the study is to offer suggestions to policy makers. These are based on our observations and analysis. We make the following recommendations. The seventh and the last suggestions are alternate in development.

1. Paper is an index of development; viewed in the context of a developing economy only production of cultural paper and newsprint should be encouraged. The four rayon units with high consumption of scarce raw material (see map on consumption of raw material) should either be closed down or converted into paper or newsprint producing units.
2. Newsprint industry should be protected by protective tariffs. Artificial dumping should be prevented.
3. Private sector PPI units which continue to use obsolete sulphate processes, should be compelled to upgrade their pulping technology so that the maximum quantity of cellulosic raw material can be utilized. Though investment in new technology is high, it will be more than offset by saving such scarce raw material.
4. The chemical load in effluents is much too high for it to be used for cultivation. Till more R and D work is carried out, it should be stopped forthwith (see Ch. 8).
5. Industrial plantations of pulpwood species in natural forests should be banned—they spell a danger to the complex forest ecosystem.
6. Farm forestry including exotic species which are a mismatch with the site, should be banned. The word banned is deliberately used as the peasantry, on whose land the trees have been planted, has only a formal legal title to their lands, but no control over the choice of species to be cultivated on it. This is a consequence of incomplete agrarian reforms in India.
7. Eco-friendly alternate raw materials which range from long-term crops like bamboo to seasonal crops like *kenaf*, hemp and their numerous native cousins have been grown by farmers for PPI. The farmers have to be given state protection for fair prices for their products (see Chs. 1 and 2).
8. The PPI should not be granted degraded forest land for captive plantations, as political pressures have been used to conveniently label forest lands as degraded land (see Chs. 2 and 6). The National

Forestry Action Plan (NFAP) looks favourably on it. NFAP is formulated by the Ministry of Forestry and LNDP in June, 1993 (*Update*, no. 32, 1995: 21).

9. Joint sector plantations should not be permitted. Foresters view them as a threat to forests (Ch. 2) and their judgement should be respected.
10. Forests should be mapped by competent authorities to allow the Forest Department to have greater control over them.
11. There is much tension in Centre-state relations particularly on the question of allocation of plan funds especially for pulpwood plantations. The earlier tension was due to gifting away forest lands on long leases for a pittance. The state foresters need resources and labour power to rejuvenate and protect the forests. Like the railways, the Forest Department should have its own separate budget.
12. Ban on felling by the industry in Karnataka has rejuvenated the bamboo, while management of bamboo forests by the people in the north-east has left the bamboo forests in good health. It's time that policy makers take note of these two facts.
13. Investigations must be carried out seriously to find who actually suppressed the FAO 1980 report and failed to convey its negative findings on *Eucalyptus tereticornis* to the various State Forest Department Corporations as well as to the PPI. CBI must also investigate more fully the newsprint scam, the dealers and officers of the Ministry of Information and Broadcasting who were involved in unfairly importing newsprint into the country to the detriment of India's newsprint industry.
14. India does not need FAO's advice. The NFAP must be scrapped. Indian specialists should review the Indian situation.
15. The state should not forget that the forest dwellers too have rights to their traditional homelands and a full right to participate in India's development and share in the gain.

NOTES

1. *Tathastu*, in Sanskrit means Amen.
2. As an experimental measure, small areas were cordoned off when bamboo seeds had sprouted after gregarious flowering in the early 1980s in the Chandrapur Forests. With no animals to trample over or browse and weed out the excess, almost every one of the sprouted seedlings seems

to have struggled for survival and growth. The struggle appears to have been so intense that the relevant patch looks like a strange tangled bed of grass sticks.

3. But the environmentally conscious foresters did not care to find out from where the James River Paper Unit within the state got its supply of raw material. It was from Canada and Brazil (An Executive, James River Paper Co. Oregon State, personal communication). James River is one of the largest paper companies in USA and its factories are spread all over the forested regions in the North-East and Western USA.
4. FAO's negative report or otherwise, government's interest in exotic pulpwood persists. Kamal Nath the Minister of Forests and Environment, in response to an unstarred question asking for information on species planted in forests in the Lok Sabha (Parliament) on 16 May 1995, answered non-committally that afforestation and choice of species is based on 'edapic' climate geomorphological and anthropogenic factors. Among the long list of species figure *Acacia auriculiformis* and *Casuarina equistifolia* as nitrogen-fixing species meant for fuelwood, small timber and fodder and eucalyptus species as economically important species. Our findings about these species do not coincide with Kamal Nath's opinion.
5. The efficacy of these medicines are at last recognized and are referred to as the 'alternate system'. The knowledge is still empirical and the pharmaco dynamics of these potent ingredients only need to be assessed for the systems to be given scientific status. But then even much of Western medicine is still empirical too.
6. *The Economic Times*, 27 July 1993, reports that UNDP has given Rs. 2.25 crore grant aid for National Forestry Action Programmes, to achieve ecological balance and sustainable development by adopting a holistic people- participatory and decentralized approach. *Business Standard* of the same date has another version. The funds, it states, will finance a range of inputs including international experts and consultants, study tours, training and equipment to be co-ordinated by a specially created cell within the Ministry of Environment and Forestry which will formulate NFAP with active involvement of state governments and special assistance from FAO. FAO officials however are puzzled as to why forestry officials in India are unfriendly. Foresters in India are not unfriendly, but do not care greatly for FAO's advice. This is understandable.
7. It is difficult to understand why the National Chemical Laboratory (NCL), Pune, has undertaken a scheme 'to coax' delayed flowering bamboo to flower within a few weeks. NCL regards this as a 'break through (which) may open up possibilities of development of short maturing and high yielding hybrids of bamboo'. NCL has also been assigned the task of micro-propagation of 'Elite' forest trees through tissue culture. The three important species are *E. tereticornis, E. camaldulensis, B. vulgaris, Selvador*

persica. This project is meant to give a fillip to the country's forestry (NABARD, *Annual Report*, 1993-4: 83). NCL has been given a grant of Rs. 126 lakh. NABARD does not recommend the species, it only disperses resources (Senior Official, NABARD, personal communication, August 1995).

REFERENCES

BIC 1994, 'Bamboo Researchers and Projects of South and South East Asia', KFRI, Peechi, India.

Chopra, P.N. (ed.) 1991, *Quit India Movement, Role of Indian Big Business, British Secret Documents*, Interprint, New Delhi.

ESRF 1969, *Pulp and Paper Prospects for 1975*, New Delhi, Ministry of Agriculture 1976, Report on National Commission of Agriculture, vol. IX, GOI, New Delhi.

FAO 1954, *World Paper Resources and Projects*, New York.

——— 1966, *Pulp and Paper Development in Africa and Near East*, vol. I, Rome.

——— 1968, *Forests and People*, Rome.

——— 1980, *Pulping and Paper Making Properties of Fast Growing Plantation Wood Species*, vols. I and II, Rome

——— 1993, *Assessing Forestry Project Impacts Issues and Strategies*, Rome.

——— 1 October 1993, *Forestry Topics Report No. 4 More than Wood*, Rome.

——— 1994, *Forest Development and Policy Dilemmas*, Part III, Rome.

——— *NFAP Update No. 32—India*, Rome.

——— 8-12 September 1995, *Travel Report—India*, Rome.

——— 1996, *Formulation, Execution and Revision of National Forestry Programmes: Basic Principles and Operational Guidelines*, Rome.

Kadambi, K. 1949, 'Ecology and Silviculture of *Dendrocalamus strictus*', *Indian Forester*, vol. LXXV, Dehradun.

Kimmins, J.P. 1987, *Forestry Ecology*, Macmillan, London and NY.

Leslie, A.J. 1994, *Technical Papers on Sustainable Forest Management*, FAO, Rome.

Saraf, V.P., R.M. Shveshwar, P.M. Meshrankar, O.K. Deb, N.S. Jaspal, B.P. Biyan 1980, 'Mesta/Kenaf as Raw Material for Kraft Pulping', *Journal of Indian Paper and Technical Association*, vol. XVII, no. 4.

Saxena, N.C. 1990, *Farm Forestry in North West India*, Ford Foundation.

Thirawat, S. 1955, 'Eucalyptus for Tropical Climates, FAQ Experience', *Indian Forester*, vol. LXXXI, no. 4.

——— 1956, 'Value of Eucalyptus in Asia', *Indian Forester*, vol. LXXXII, no. 12.

Tushu, N.K. and N.S.R. Murthy 1991, 'Mesta, A Promising Indian Fiberous Raw Material', *Indian Pulp and Paper Technical Association,* vol. III, no. 3 Saharanpur, India.

Troupe, R.S. 1921, *Silviculture of Indian Trees*, vol. III, International Book Distributors, Dehradun.

Von Monroy, J.A. 1961, *Report to Government of India on Integration of Forests and Forest Industries,* FAO.

Index